Compendium of Plant Genomes

Series Editor

Chittaranjan Kole, President, International Climate Resilient Crop Genomics Consortium (ICRCGC), President, International Phytomedomics & Nutriomics Consortium (IPNC) and President, Genome India International (GII), Kolkata, India

Whole-genome sequencing is at the cutting edge of life sciences in the new millennium. Since the first genome sequencing of the model plant *Arabidopsis thaliana* in 2000, whole genomes of about 100 plant species have been sequenced and genome sequences of several other plants are in the pipeline. Research publications on these genome initiatives are scattered on dedicated web sites and in journals with all too brief descriptions. The individual volumes elucidate the background history of the national and international genome initiatives; public and private partners involved; strategies and genomic resources and tools utilized; enumeration on the sequences and their assembly; repetitive sequences; gene annotation and genome duplication. In addition, synteny with other sequences, comparison of gene families and most importantly potential of the genome sequence information for gene pool characterization and genetic improvement of crop plants are described.

More information about this series at https://link.springer.com/bookseries/11805

Malali Gowda • H. B. Mahesh •
Chittaranjan Kole
Editors

The Sandalwood Genome

Springer

Editors
Malali Gowda
Bengaluru Genomics Center
Bangalore, India

H. B. Mahesh
Genetics and Plant Breeding
University of Agricultural Sciences
Bangalore, India

Chittaranjan Kole
International Climate Resilient Crop
Genomics Consortium (ICRCGC)
International Phytomedomics
and Nutriomics Consortium (IPNC)
Genome India International (GII)
Kolkata, India

ICAR-National Institute for Plant
Biotechnology
Pusa, New Delhi, India

ISSN 2199-4781 ISSN 2199-479X (electronic)
Compendium of Plant Genomes
ISBN 978-3-030-93393-7 ISBN 978-3-030-93394-4 (eBook)
https://doi.org/10.1007/978-3-030-93394-4

This Springer imprint is published by the registered company Springer Nature Switzerland AG
The registered company address is: Gewerbestrasse 11, 6330 Cham, Switzerland

This book series is dedicated to my wife Phullara and our children Sourav and Devleena

Chittaranjan Kole

Preface to the Series

Genome sequencing has emerged as the leading discipline in the plant sciences coinciding with the start of the new century. For much of the twentieth century, plant geneticists were only successful in delineating putative chromosomal location, function, and changes in genes indirectly through the use of a number of "markers" physically linked to them. These included visible or morphological, cytological, protein, and molecular or DNA markers. Among them, the first DNA marker, the RFLPs, introduced a revolutionary change in plant genetics and breeding in the mid-1980s, mainly because of their infinite number and thus potential to cover maximum chromosomal regions, phenotypic neutrality, absence of epistasis, and codominant nature. An array of other hybridization-based markers, PCR-based markers, and markers based on both facilitated construction of genetic linkage maps, mapping of genes controlling simply inherited traits, and even gene clusters (QTLs) controlling polygenic traits in a large number of model and crop plants. During this period, a number of new mapping populations beyond F2 were utilized and a number of computer programs were developed for map construction, mapping of genes, and for mapping of polygenic clusters or QTLs. Molecular markers were also used in the studies of evolution and phylogenetic relationship, genetic diversity, DNA fingerprinting, and map-based cloning. Markers tightly linked to the genes were used in crop improvement employing the so-called marker-assisted selection. These strategies of molecular genetic mapping and molecular breeding made a spectacular impact during the last one and a half decades of the twentieth century. But still they remained "indirect" approaches for elucidation and utilization of plant genomes since much of the chromosomes remained unknown and the complete chemical depiction of them was yet to be unraveled.

Physical mapping of genomes was the obvious consequence that facilitated the development of the "genomic resources" including BAC and YAC libraries to develop physical maps in some plant genomes. Subsequently, integrated genetic–physical maps were also developed in many plants. This led to the concept of structural genomics. Later on, emphasis was laid on EST and transcriptome analysis to decipher the function of the active gene sequences leading to another concept defined as functional genomics. The advent of techniques of bacteriophage gene and DNA sequencing in the 1970s was extended to facilitate sequencing of these genomic resources in the last decade of the twentieth century.

As expected, sequencing of chromosomal regions would have led to too much data to store, characterize, and utilize with the-then available computer software could handle. But the development of information technology made the life of biologists easier by leading to a swift and sweet marriage of biology and informatics, and a new subject was born—bioinformatics.

Thus, the evolution of the concepts, strategies, and tools of sequencing and bioinformatics reinforced the subject of genomics—structural and functional. Today, genome sequencing has traveled much beyond biology and involves biophysics, biochemistry, and bioinformatics!

Thanks to the efforts of both public and private agencies, genome sequencing strategies are evolving very fast, leading to cheaper, quicker, and automated techniques right from clone-by-clone and whole-genome shotgun approaches to a succession of second-generation sequencing methods. The development of software of different generations facilitated this genome sequencing. At the same time, newer concepts and strategies were emerging to handle sequencing of the complex genomes, particularly the polyploids.

It became a reality to chemically—and so directly—define plant genomes, popularly called whole-genome sequencing or simply genome sequencing.

The history of plant genome sequencing will always cite the sequencing of the genome of the model plant Arabidopsis thaliana in 2000 that was followed by sequencing the genome of the crop and model plant rice in 2002. Since then, the number of sequenced genomes of higher plants has been increasing exponentially, mainly due to the development of cheaper and quicker genomic techniques and, most importantly, the development of collaborative platforms such as national and international consortia involving partners from public and/or private agencies.

As I write this preface for the first volume of the new series "Compendium of Plant Genomes," a net search tells me that complete or nearly complete whole-genome sequencing of 45 crop plants, eight crop and model plants, eight model plants, 15 crop progenitors and relatives, and three basal plants is accomplished, the majority of which are in the public domain. This means that we nowadays know many of our model and crop plants chemically, i.e., directly, and we may depict them and utilize them precisely better than ever. Genome sequencing has covered all groups of crop plants. Hence, information on the precise depiction of plant genomes and the scope of their utilization are growing rapidly every day. However, the information is scattered in research articles and review papers in journals and dedicated Web pages of the consortia and databases. There is no compilation of plant genomes and the opportunity of using the information in sequence-assisted breeding or further genomic studies. This is the underlying rationale for starting this book series, with each volume dedicated to a particular plant.

Plant genome science has emerged as an important subject in academia, and the present compendium of plant genomes will be highly useful to both students and teaching faculties. Most importantly, research scientists involved in genomics research will have access to systematic deliberations on the plant genomes of their interest. Elucidation of plant genomes is of interest not only for the geneticists and breeders, but also for practitioners of an array of plant science disciplines, such as taxonomy, evolution, cytology, physiology, pathology, entomology, nematology, crop production, biochemistry, and obviously bioinformatics. It must be mentioned that information regarding each plant genome is ever-growing. The contents of the volumes of this compendium are, therefore, focusing on the basic aspects of the genomes and their utility. They include information on the academic and/or economic importance of the plants, description of their genomes from a molecular genetic and cytogenetic point of view, and the genomic resources developed. Detailed deliberations focus on the background history of the national and international genome initiatives, public and private partners involved, strategies and genomic resources and tools utilized, enumeration on the sequences and their assembly, repetitive sequences, gene annotation, and genome duplication. In addition, synteny with other sequences, comparison of gene families, and, most importantly, the potential of the genome sequence information for gene pool characterization through genotyping by sequencing (GBS) and genetic improvement of crop plants have been described. As expected, there is a lot of variation of these topics in the volumes based on the information available on the crop, model, or reference plants.

I must confess that as the series editor, it has been a daunting task for me to work on such a huge and broad knowledge base that spans so many diverse plant species. However, pioneering scientists with lifetime experience and expertise on the particular crops did excellent jobs editing the respective volumes. I myself have been a small science worker on plant genomes since the mid-1980s and that provided me the opportunity to personally know several stalwarts of plant genomics from all over the globe. Most, if not all, of the volume editors are my longtime friends and colleagues. It has been highly comfortable and enriching for me to work with them on this book series. To be honest, while working on this series I have been and will remain a student first, a science worker second, and a series editor last. And I must express my gratitude to the volume editors and the chapter authors for providing me the opportunity to work with them on this compendium.

I also wish to mention here my thanks and gratitude to the Springer staff, particularly Dr. Christina Eckey and Dr. Jutta Lindenborn for the earlier set of volumes and presently Ing. Zuzana Bernhart for all their timely help and support.

I always had to set aside additional hours to edit books beside my professional and personal commitments—hours I could and should have given to my wife, Phullara, and our kids, Sourav and Devleena. I must mention that they not only allowed me the freedom to take away those hours from them but also offered their support in the editing job itself. I am really not sure whether my dedication of this compendium to them will suffice to do justice to their sacrifices for the interest of science and the science community.

New Delhi, India Chittaranjan Kole

Preface

Indian sandalwood is one of the most valuable trees in the world. The essential oil extracted from heartwood has gained a significant place in perfumery industries. Unfortunately, anthropogenic activity has caused a gradual decrease in the production of sandalwood in India. Recent data on production of sandalwood in India showed a declining trend and it is considered as endangered due to over-harvesting of roots of this tree. There is a limited research done to understand the genetic diversity of sandalwood using genome-wide markers. In addition, lack of genomic resources for this highly prized tree, was an impediment for the diversity studies and its conservation in the natural habitat. The genomic studies in sandalwood have primarily focused on understanding the molecular mechanism governing santalol oil content. However, understanding the sandalwood at whole-genome level was unattended till recent past. Our motivation is to compile the genomic, transcriptomic, and proteomic information into a book. This book can serve as a ready reckoner for sandalwood researchers. Recent advances in genome sequencing and transcriptome profiling revealed a greater insight on the transformation of sandal with gene(s) resistant to various sandal diseases and for metabolic engineering to increase or modify the essential oil yield. The characterization and functional annotation of genes responsible for the production of essential oil in sandalwood, is one of the major areas of focus due to its economic importance in preparation of various perfumes and cosmetics.

Whole genome sequence of different sandalwood ectypes, its comparative studies, and multiomic information provided in this book holds wide implication for future research and has potential to reach out to the global scientific community. Very importantly, the combination of information from multi-omics resources will enable trait-specific breeding and conservation of genetic resources of *S. album*. The molecular understanding of the sesquiterpene pathway will have industrial implications through biomarker discovery and metabolic engineering in microbial system for santalol production.

Bangalore, India	H. B. Mahesh
Bangalore, India	Malali Gowda
Kolkata, India	Chittaranjan Kole

Contents

Contributors

Tabassum Ishrath Fathima Centre for Traditional Knowledge & Informatics, The University of Trans-Disciplinary Health Sciences and Technology (TDU), Bengaluru, India

Modhumita Ghosh Dasgupta Institute of Forest Genetics and Tree Breeding, R.S. Puram, Coimbatore, India

Malali Gowda The University of Trans-Disciplinary Health Sciences and Technology (TDU), Jarakabande Kaval, Bengaluru, India;
Bengaluru Genomics Center, Yelahanka, Bengaluru, India

Chittaranjan Kole ICAR-National Research Center on Plant Biotechnology, Pusa, New Delhi, India

Chandramouli K. Madhuvanthi Institute of Forest Genetics and Tree Breeding, R.S. Puram, Coimbatore, India

H. B. Mahesh Department of Genetics and Plant Breeding, College of Agriculture, University of Agricultural Sciences, Mandya, Bengaluru, India

K. G. Manasa Department of Genetics and Plant Breeding, College of Agriculture, Mandya, University of Agricultural Sciences, Bangalore, India

T. N. Manohara Silviculture and Forest Management Division, Institute of Wood Science and Technology, Malleswaram, Bengaluru, India

Eswaran Muthulakshmi Institute of Forest Genetics and Tree Breeding, R.S. Puram, Coimbatore, India

Dinesh A. Nagegowda Molecular Plant Biology and Biotechnology Lab, CSIR-Central Institute of Medicinal and Aromatic Plants, Bengaluru, India;
Academy of Scientific and Innovative Research (AcSIR), Ghaziabad, India

S. N. Venugopalan Nair Centre for Traditional Knowledge & Informatics, The University of Trans-Disciplinary Health Sciences and Technology (TDU), Bengaluru, India

Ananth K. Narayanan Molecular Plant Biology and Biotechnology Lab, CSIR-Central Institute of Medicinal and Aromatic Plants, Bengaluru, India;
Academy of Scientific and Innovative Research (AcSIR), Ghaziabad, India

Shilpa Naveen Centre for Traditional Knowledge & Informatics, The University of Trans-Disciplinary Health Sciences and Technology (TDU), Bengaluru, India

Thottethodi Subrahmanya Keshava Prasad Center for Systems Biology and Molecular Medicine, Yenepoya Research Center, Yenepoya (Deemed to be University), Mangalore, India

M. K. Prasannakumar Department of Plant Pathology, College of Agriculture, GKVK, University of Agricultural Sciences, Bangalore, India

N. R. Raghavendra Department of Genetics and Plant Breeding, College of Agriculture, Mandya, University Agricultural Sciences, Bangalore, India

C. Sandeep Silviculture and Forest Management Division, Institute of Wood Science and Technology, Malleswaram, Bengaluru, India

Meghana Deepak Shirke Department of Biotechnology, KTHM College, Nashik, India

Sudharshan SJ Molecular Plant Biology and Biotechnology Lab, CSIR-Central Institute of Medicinal and Aromatic Plants, Bengaluru, India

Pratigya Subba Center for Systems Biology and Molecular Medicine, Yenepoya Research Center, Yenepoya (Deemed to be University), Mangalore, India

Abbreviations

AACT	Acetyl-CoAC-acetyl Transferase
ANI	Average Nucleotide Identity
bRPLC	basic Reverse-phase Liquid Chromatography
BUSCO	Benchmarking Universal Single-Copy Orthologs
CEGMA	Core Eukaryotic Genes Mapping Approach
CEGs	Core Eukaryote Genes
CPP	Chrysanthemyl Diphosphate
CPR	Cytochrome P450 Reductase
ddRADSeq	Double Digested Restriction Associated DNA Sequencing
DMAPP	Dimethylallyl Diphosphate
DXP	1-deoxy-D-xylulose 5-phosphate
ER	Endoplasmic Reticulum
FDA	Food and Drug Administration
FPP	Farnesyl Diphosphate
G3P	D-glyceraldehyde 3-phosphate
GAL	Galactose Inducible Promoter
GBS	Genotyping-by-sequencing
GGPP	Geranylgeranyl Pyrophosphate
GGPPS	GGPP synthase
GPP	Geranyl Diphosphate
GPPS	GPP synthase
HMG-CoA	3-hydroxy-3-methylglutaryl coenzyme A
HMGR	3-hydroxy-3-methylglutaryl-CoA reductase
HMGS	3-hydroxy-3-methyglutary-CoA synthase
HW	Heartwood
iBAQ	Intensity-Based Absolute Quantification
InDels	Insertions and Deletions
IPP	Isopentenyl Diphosphate
ISO	International Organization for Standardization
ISSR	Inter Simple Sequence Repeat
IUCN	International Union for Conservation of Nature
KOG	EuKaryotic Orthologous Groups
LPP	Lavandulyl Diphosphate
LSC	Large Single Copy
LSU	Large Subunit
MAS	Marker Assisted Selection

Mbp	Million base pair
MeJA	Methyl Jasmonate
MEP	Methylerythritol Phosphate
MEV	Mevalonate
MP	Mate-pair
NGS	Next Generation Sequencing
NNPP	Nerylneryldiphosphate
PE	Paired-end
PGM	Phosphoglucomutase
RAPD	Random Amplified Polymorphic DNA
RET	Rare Endangered Threatened
RFLP	Restriction Fragment Length Polymorphism
SA	Salicylic Acid
SNPs	Single Nucleotide Polymorphisms
SQ	Squalene
SQS	Squalene Synthase
SRA	Sequence Read Archive
SSD	Sandal Spike Disease
SSR	Simple Sequence Repeat
SSU	Small Subunit
SVs	Structural Variations
SW	Sapwood
TGA	Therapeutic Goods Administration
TPS	Terpene Synthases
Ts	Transition
Tv	Transversions
TZ	Transition Zone
WGS	Whole Genome Sequencing

Sandalwood: The King of Aroma in Plant Kingdom

1

H. B. Mahesh and Malali Gowda

Abstract

The *Santalum* is the most important genus under the *Santalaceae* family. The Indian sandalwood, *Santalum album*, is the only economically important tree species of this family traded for its fragrant essential oil. *Santalum album* is indigenous to the tropical regions of the Indian peninsula (Deccan Plateau), eastern Indonesia, and northern Australia. It had declared as the Royal tree by Tipu Sultan in 1792 and has an inextricable link with Indian culture and heritage. *Santalum album* is an evergreen, slow growing, hemiparasitic tree with a diploid chromosome number of $2n = 20$. Limited genomic resources in the public databases for sandalwood triggered enumeration of three genomes from different research groups. Many integrated omics technologies have helped to characterize genes encoding for santalol biosynthesis.

H. B. Mahesh (✉)
Department of Genetics and Plant Breeding, College of Agriculture, University of Agricultural Sciences, Mandya, Bengaluru 560065, India

M. Gowda
The University of Trans-Disciplinary Health Sciences and Technology (TDU), Jarakabande Kaval, Bengaluru 560065, India

M. Gowda
Bengaluru Genomics Center, Yelahanka, Bangalore 560064, India

1.1 Introduction

Sandalwood is an important evergreen, slow growing, hemiparasitic tree distributed throughout the tropical and temperate regions of India, Indonesia, Australia, and Pacific Islands (Harbaugh and Baldwin 2007). This tree is associated with Indian culture and is the most expensive wood in the world after African blackwood (*Dalbergia melanoxylon*). The sandalwood oil has gained international popularity for its high quality. Nearly 90% of natural sandalwood population was concentrated in southern Karnataka and northern part of Tamil Nadu. Excessive harvesting and smuggling of sandalwood without replenishing the tree population in the ecosystem has put a sandalwood industry in shortage of supply of heartwood and led to soaring market prices. In addition, dreaded spike disease is a major biotic stress seriously infecting natural sandalwood habitats in Karnataka and Kerala. All these factors resulted in sandalwood being classified as "vulnerable" by the International Union for Conservation of Nature (IUCN) in 1998. Now, sandalwood tree biologists should step forward to effectively utilize the available genomic resources and molecular biology tools to improve this highly valued tree of the plant kingdom.

M. Gowda et al. (eds.), *The Sandalwood Genome*, Compendium of Plant Genomes,
https://doi.org/10.1007/978-3-030-93394-4_1

1.2 Taxonomic Classification

Sandalwood belongs to genus *Santalum* and it consists of 15 extant species.

Kingdom:	Plantae
Subkingdom:	Viridiplantae
Division:	Tracheophyta
Subdivision:	Spermatophytina
Class:	Magnoliopsida
Superorder:	Santalanae
Order:	Santalales
Family:	Santalaceae
Genus:	Santalum
Species:	*Santalum album* L.
Common name:	Sandalwood

1.3 Botanical Description

Santalum album is a small evergreen tree that grows to a height of 20–30 m with slender drooping branches (Fig. 1.1a). Fully matured tree attains a girth up to 2.4 m and may live more than 100 years of age. Bark is tight, dark brown, reddish, dark grey to nearly black, smooth in young trees, rough with deep vertical cracks in older trees, red inside (Fig. 1.1b). Leaves are thin, leathery, opposite in orientation on branch, ovate or ovate elliptical in shape, upper surface is glabrous (hairless) and shining green, lower surface is glaucous and slightly paler (Fig. 1.1c). The leaf tip is pointed having reticulate venation. Flowers are purplish-brown, unscented, and small in size (Fig. 1.1d). Flowers borne in axillary or terminal cymose panicles exhibit acropetal succession (Fig. 1.1e). The ovary is semi-inferior and unilocular with three ovules. Flowering is seen twice a year during March–April and September–October. Usually trees start flowering from three years after planting. Fruit is globose, fleshy drupe, purple to black when ripe, almost stalkless, smooth and single seeded (Fig. 1.1f). Shapes of the fruit vary from round to oblong. Seeds are naked lacking testa. The birds which feed on outer fleshy pericarp of seeds and seeds are easily dispersed by them which help in propagation in the natural habitat.

1.4 Cytology

The somatic chromosome number of *S. album* is $2n = 20$ (Goldblatt and Johnson 2000; Harbaugh 2008; Rao 1942; Xin-Hua et al. 2010) and estimated haploid DNA content is 0.21 pg (Mahesh et al. 2018). A mixoploidy was also reported which is having cells of $2n = 20$ and $2n = 40$ (Xin-Hua et al. 2010). The chromosomes are short, length ranging from 1.16 to 2.33 μm. Majority of the chromosomes are having centromere in median position and few chromosomes possess submedian centromere. Secondary constriction is seen on long arm of third pair of chromosomes.

1.5 Distribution/Ecology

Santalum album is indigenous to the tropical region of the Indian peninsula (Deccan Plateau), eastern Indonesia, and northern Australia. There are 16 recognized species of sandalwood spread over in the world; broadly they can be grouped as East Indian sandalwood (*Santalum album*), Australian sandalwood (*Santalum spicatum*), Hawaiian sandalwood, and sandalwood of the Pacific Islands. Many of these species are threatened due to over-exploitation for their highly treasured essential oils. East Indian sandalwood, *S. album* is the queen among sandalwood species attributed for its high yielding superior quality essential oil which is essential for perfumery and related industries. In India, *S. album* is seen in Karnataka, Tamil Nadu, Kerala, Andhra Pradesh, Madhya Pradesh, Chhattisgarh, Maharashtra, Orissa, Rajasthan, West Bengal, Manipur, Gujarat, Uttar Pradesh, Uttarakhand, Himachal Pradesh, Jammu & Kashmir, Assam, and Punjab. The major distribution is seen at the edge of the Western Ghats range, Karnataka, and Tamilnadu. It is opined that 200 km radius of Bangalore city is found to be the main zone of

Fig. 1.1 Various parts of Sandalwood tree

natural distribution of sandalwood (Rai 1990). This species exhibits profuse/luxuriant growth under high rainfall areas but heartwood formation is negligible or absent. Sandalwood can thrive under different kinds of soils like sand, clay, laterite, loam, and black cotton soil with pH ranging from 7 to 8.5.

1.6 Habit

Sandalwood is a hemi root parasite and can parasitize over 300 species. The bell-shaped mature haustoria establishes connections with host by penetrating its peg into host cell. It is enabled by direct pressure and secretion of cell-wall degrading enzymes. It is a phytohormone-mediated regulation process where auxin signaling pathway is responsible for haustorial initiation, subsequently genes related to cytokinin and gibberellin biosynthesis have a major role in haustoria development (Zhang et al. 2012, 2015, 2017). After successful connection with xylem tissue of the host root, acquisition of nutrients takes place for growth and development.

1.7 Uses of Sandalwood

Indian sandalwood (*S. album*) is considered to be one of the most valuable sandalwood species. Heartwood is the most valuable part of sandal wood tree and fragrant essential oil is extracted from the same. The value of sandalwood mainly depends on the proportion of heartwood and on the concentration and quality of essential oil extracted from the heartwood. It is one of the first traded item from India to other countries. Indian sandalwood usually yields around 3–8% essential oil. Over 85% of *S. album* essential oil consists of sesquiterpenoids, including sesquiterpene alcohols (α-santalol, β-santalol, *trans*-α-bergamotol, and epi-β-santalol) and their corresponding sesquiterpenes (α-santalene, β-santalene, *trans*-α-bergamotene, and epi-β-santalene). In Hinduism, Buddhism, and other religions, sandalwood is used for burning dead bodies during cremation. The sapwood paste and oil are used as coolants. Sandalwood oil is mainly used in perfumery, medicine, cosmetic, and soap industries. Sandalwood is also used for carving. Other uses are detailed in subsequent chapters of this book.

1.8 Availability of Genomic Resources

Santalum album being the most valuable tree, genomic resource available for this tree is very limited. The NCBI database search yielded the following dataset with "*Santalum album*" keyword as of 10/05/2021. Currently, there are three genomes available in the NCBI database (NCBI Genome accession numbers: LOCJ00000000, WFLO00000000, and NXEK00000000) which have immense importance in the sandalwood improvement (Dasgupta et al. 2019; Mahesh et al. 2018). In addition, RNA sequencing of global transcriptome of leaf, cold stress tolerance, haustorium development, and synthesis of essential oil in heartwood has added value to understand gene expression pattern (Celedon et al. 2016; Diaz-Chavez et al. 2013; Jones et al. 2011; Moniodis et al. 2015; Srivastava et al. 2015; Zhang et al. 2012, 2015, 2017). Conventional cloning or subtractive hybridization-derived process enabled to deposit 682 full or partial gene sequences. Both DNA and RNA sequence data have been deposited in sequence read archive (SRA) database from several researchers.

Database	Number
Assembly	3
Bioproject	14
Biosample	37
Genome	1
Nucleotide	682
SRA	34
Gene	124
Geo datasets	46
Popset	210
Protein	1015

References

Celedon JM, Chiang A, Yuen M, Diaz-Chavez ML, Madilao LL et al (2016) Heartwood-specific transcriptome and metabolite signatures of tropical sandalwood (*Santalum album*) reveal the final step of (Z)-santalol fragrance biosynthesis. Plant J 86(4):289–299

Dasgupta MG, Ulaganathan K, Dev SA, Balakrishnan S (2019) Draft genome of *Santalum album* L. Provides genomic resources for accelerated trait improvement. Tree Genet Genom 15(3):1–15

Diaz-Chavez ML, Moniodis J, Madilao LL, Jancsik S, Keeling CI et al. (2013) Biosynthesis of sandalwood oil: *Santalum album* CYP76F cytochromes P450 produce santalols and bergamotol. PLoS One 8(9):e75053

Goldblatt P, Johnson D (2000) Index to plant chromosome numbers 1996–1997. Monogr Syst Bot 81:188

Harbaugh DT (2008) Polyploid and hybrid origins of Pacific island sandalwoods (Santalum, Santalaceae) inferred from low-copy nuclear and flow cytometry data. Int J Plant Sci 169(5):677–685

Harbaugh DT, Baldwin BG (2007) Phylogeny and biogeography of the sandalwoods (Santalum, Santalaceae): repeated dispersals throughout the Pacific. Am J Bot 94(6):1028–1040

Jones CG, Moniodis J, Zulak KG, Scaffidi A, Plummer JA et al (2011) Sandalwood fragrance biosynthesis involves sesquiterpene synthases of both the terpene synthase (TPS)-a and TPS-b subfamilies, including santalene synthases. J Biol Chem 286(20):17445–17454

Mahesh HB, Subba P, Advani J, Shirke MD, Loganathan RM et al (2018) Multi-omics driven assembly and annotation of the sandalwood (*Santalum album*) genome. Plant Physiol 176(4):2772–2788

Moniodis J, Jones CG, Barbour EL, Plummer JA, Ghisalberti EL et al (2015) The transcriptome of sesquiterpenoid biosynthesis in heartwood xylem of Western Australian sandalwood (*Santalum spicatum*). Phytochemistry 113:79–86

Rao L (1942) Studies in the Santalaceae. Ann Bot 6 (21):151–175

Srivastava PL, Daramwar PP, Krithika R, Pandreka A, Shankar SS et al (2015) Functional characterization of novel sesquiterpene synthases from Indian sandalwood, *Santalum album*. Sci Rep 5:1–12

Rai SN (1990) Status and cultivation of sandalwood in India. In: Hamilton, Lawrence; Conrad, C. Eugene, technical coordinators. Proceedings of the symposium on sandalwood in the Pacific; April 9–11, 1990; Honolulu, Hawaii. Gen. Tech. Rep. PSW-GTR-122. Berkeley, CA: Pacific Southwest Research Station, Forest Service, US Department of Agriculture, pp 66–71

Xin-Hua Z, da Silva JAT, Ma G-H (2010) Karyotype analysis of *Santalum album* L. Caryologia 63(2):142–148

Zhang X, Berkowitz O, da Silva JAT, Zhang M, Ma G et al (2015) RNA-Seq analysis identifies key genes associated with haustorial development in the root hemiparasite *Santalum album*. Front Plant Sci 6:661

Zhang X, da Silva JAT, Duan J, Deng R, Xu X et al. (2012) Endogenous hormone levels and anatomical characters of haustoria in *Santalum album* L. seedlings before and after attachment to the host. J Plant Physiol 169(9):859–866

Zhang X, da Silva JAT, Niu M, Li M, He C et al. (2017) Physiological and transcriptomic analyses reveal a response mechanism to cold stress in Santalum album L. leaves. Sci Reports 7(1):1–18

2 Origin, Tradition and Cultural Importance of Sandalwood in India and Its Role in Conservation of Natural Populations

C. Sandeep and T. N. Manohara

Abstract

Trees of ancient India had divine beliefs, values and religious significance which influenced where, why and which trees were to be planted, protected and cut. Many tree species have pre-eminent role in Indian Culture, tradition and are regarded as trees of eternal value in *Sanatana Dharma*. Oriental and Daoic religions of Asia provided trees a sacred place. Indian sandalwood (*Santalum album* Linn) is an epitome of elegance and the rarest jewel of the plant kingdom in the realms of nature. It is the most valued timber in the whole of Asia. It is a commodity preferred by most nations for over 2,000 years due to its unique set of natural properties. The heartwood of *S. album* is highly priced for its aromatic oil in international markets. Due to illegal felling and over-exploitation, this species is now enlisted in the vulnerable category of IUCN Red List. It is commonly known by its Sanskrit name 'Chandana' in peninsular India. Its noteworthy association with the social rituals of Hinduism is an example of highly advanced civilisations of ancient India that needs a special mention and adoration. It equally holds a special place in all the major religions of India. Its divine association with Hinduism is documented in the written records, hymns and oral incantations of ancient literatures. In spite of its commercial legacies and cultural heritage, the sandalwood improvement programme in India is limited. Owing to its unique importance there is an urgent need to conserve this species. Current study enthrals the beliefs, its importance and significance of traditional knowledge in effective conservation.

2.1 Introduction

White sandalwood (*S. album* L. Family: Santalaceae) is known as 'Dollar parasite' (Krishnappa 1972; Durairaj and Kamraj 2013) and its oil as 'Queen of oils' (Radomiljac et al. 1998). In spite of the fact that varying number of species have been reported by several researchers, only 16 important species have been categorized. A single species of *Santalum* (*S. album*) inhibits the Indian subcontinent and Asian region (Ramaswamy 1956; Fox 2000). Indian sandalwood is a medium sized evergreen tree of dry deciduous forests that can grow up to a height of 18 to 20 m and attain a girth of 2–2.5 m (De Candolle 1857; Sen Sarma 1982). This hardy species can thrive from sea level to 5000 m altitude. It is an obligate hemi-parasite that can associate with more than 300–350 plants species (Rama Rao 1911;

C. Sandeep (✉) · T. N. Manohara
Silviculture and Forest Management Division, Institute of Wood Science and Technology, 18th Cross, Malleswaram, Bengaluru, India

M. Gowda et al. (eds.), *The Sandalwood Genome*, Compendium of Plant Genomes,
https://doi.org/10.1007/978-3-030-93394-4_2

Annapurna et al. 2006) and adapts well in different agro-climatic conditions (Rama Rao 1908; Troup 1921; Annapurna et al. 2006). All known species of '*Santalum*' are hemi-root parasites that absorb nutrients from its host through specialized root connections known as 'haustoria' (Annapurna et al. 2006).

Leaves exhibit variation and six morphotypes have been identified (De Candolle 1857). Stem is brownish and turns hard in the later stages of its life cycle. Sapwood is white in colour which is used to make incense sticks and turnery materials. The heartwood is bitter antipyretic, cooling exhilarating and moderately hard, heavy durable porous wood which seasons slowly. It delivers a pleasant, strong smell and is yellow or brownish in appearance. The tree is valued for the following features (1) the volume of heartwood (2) the concentration of oil and (3) quality of its heartwood oil (Doran et al. 2005; Brand et al. 2007). Since earlier times sandalwood has been sorted into 18 classes (Venkatesan 1980) and its economic worth is estimated accordingly. The essential oil obtained from its heartwood is highly valued for its sweet and lasting bouquet. It has spicy, warm wood notes and has non-varying compositions with fixative properties. It is very rich in isomeric alcohols α- and β- santalol. Sandalwood oil is one of the oldest perfumery materials in the world. India was once a world leader in sandalwood production, but in the current scenario extraction of sandalwood from forest and its export to foreign countries is completely banned. From the last two centuries sandalwood policies not only contributed for 'Sandalwood monopoly and Veerappan clan' but also failed to protect the natural populations (Viswanath et al 2009). Prevailing unemployment and poverty have made sandalwood exploitation primary means for income generation in sandalwood grown regions.

It is estimated that the present market requirement is at around 28,000 tons per year, which equates to a market value of around USD 3 billion. About 67% of this revenue is generated from heartwood sales, while the rest is from the sale of Indian sandalwood oil. The Middle East, USA and France are the largest importers of Sandalwood oil. To meet the global demand Australia, Cambodia, China, Costa Rica, Indonesia, Pacific Islands, Sri Lanka, Thailand and other countries have started bulk sandalwood plantations (Venkatesha 2011). Management of sandalwood in its area of distribution (Karnataka and Tamil Nadu) needs critical attention as 90% of natural population is lost over the last two decades and therefore require rehabilitation as an endangered species (Sandeep and Manohara 2019; Sandeep et al 2020).

2.2 *Santalum Album* in Indian Culture and Tradition

2.2.1 Origin of *S. Album* and Significance of Its Heartwood

Forests play a major role in enhancing the quality of environment and preserve living resources. N. I. Vavilov proposed the concept of 'centres of origin' for the first time based on biological diversity and the centres of origin which may also be called the 'centres of diversity'. India is one among 12 Vavilovian mega centres of diversity known as Indian (Hindustan) Centre for Origin of Crops and Plant diversity. India is considered as one of the species-rich nations which shows high endemism and has four global biodiversity hotspots namely Eastern Himalaya, Indo-Burma, Western Ghats and Sri Lanka and Sundaland. Edaphic factors, climatic as well as topographic conditions and years of geological stability have given rise to a wide range of ecosystems. Tropical forests such as Western Ghats have huge diversity of tree species and this variation can act as the basis for selection and improvement of forest trees (Kemp et al. 1993). India accounts for 4 to 5% of all known plant (45,500) species on earth (NBAP 2008). Apart from Himalayan forests the biodiversity of tropical forests is found in Southern Peninsula.

Since time immemorial Indian sandalwood has been an inseparable part of human civilisation and a favoured item of craftsmen (Kushalapa 1998). Indian sandalwood has unique association with Indian heritage, culture and tradition (Fox 2000). It is known as Chandana (Sanskrit), Chandan (Hindi), white sandalwood (English), Bai-tan-xiang (Chinese), Savtador or Santalon (Greek), Bois de santal (French), Sandelholz (German), Byaku-dan (Japanese), Sandalo (Spainish and Italian), Sadalo branco (Portugese), Cendana (Indonesia), Sandal abiaz (Arabic), Chendana (Malayasia) and so on. Sandalwood has nearly fifteen different names in various Indian languages, Chandan (Hindi, Bengali, Punjabi and Marathi), Srigandha (Kannada), Sandhanam (Tamil), Chandanam (Malayalam), Sukesh (Gujarati), Sukhad (Sindhi), Valgaka or Chandono (Odhisi), Boga Chandon (Assamese), Chachandan (Manipuri), Bhadrashri or Chandanamu (Telugu), Safaid (Urdu) and so on. It is frequently mentioned in the ancient Sanskrit writings, some of which are even before the Christian era (Anonymous 1972). Its devoted retrospective with Indian culture and heritage has made it the most precious and valuable timber among Indian trees. It is regarded as the symbol of purity; its paste has high spiritual significance especially in ritualistic practices. Largest market of sandalwood are the products meant for religious, ritual and ceremonial purposes. Sandalwood has strong spiritual association with Indian tradition and its heartwood is mostly used in all aspects of human life.

S. album is considered as the 'State tree of Karnataka' and Karnataka is known as the land of origin of sandalwood (*Gandhada Nadu*). Likewise, the area in and around 'Mysuru' is eulogized as the shrine of sandalwood (*Gandhada Gudi*) (Adkoli 1977; Fox 2000). Sandalwood is believed to be probably originated from the South Western Ghats of India, especially from Coorg (Kodagu) in Karnataka. Ramaswamy (1956) reported that India is the origin of *S. album* and his observations were based on the recorded data of its occurrence in India which was of more than 2300 years old. It can be said that from the thirteenth to eighteenth century *S. album* was a significant economic resource of Deccan India (Ganeshaiah et al. 2007). Now it is a well-known fact that *S. album* is indigenous to the Deccan plateau which is getting naturalized in other parts of India and elsewhere (Sandeep et al. 2016). Over the past decade Indian sandalwood has been introduced to northern, central (Srinivasan et al. 1992), north-western (Das and Tah 2013) and north-eastern India (Babita et al. 2016). For countless centuries Indian sandalwood was used in the production of attars and was traded to middle eastern countries. Attar is a blend of sandalwood oil and flower oil, such as rose petal, jasmine, kewda, etc., and the quality of attar depends on the concentration of flower effervescence within sandalwood oil. Rare availability of sufficient quantity of sandalwood has made oil industries shift towards alternatives of Western Australian, Fiji, New Caledonia and South African *Santalum* species. Among all the species *Santalum*, *S. album* is the most widely preferred species in international and national markets.

Heartwood of sandalwood is the finest wood for woodworking, carving and turning due to its even texture (straight, closely spaced grains and uniform fibres) with an oily feel upon touch. From the heartwood; garlands, figures and statues, carved boxes, jewellery boxes, small utility boxes, chariots, elephants, camel, photo frames, fans and fan handles, letter-openers, pen holders, combs, card cases and countless number of other fancy items are prepared and exported to foreign countries (Pearson and Brown 1932). More than a decade of research and overwhelming demand it has been proved that the Indian sandalwood is the sole species yielding high quality heartwood and oil.

2.2.2 Importance of Indian Sandalwood Oil

The Sanskrit manuscripts that are more than 4000 years old cite sandalwood oil as the oldest known perfume material. According to its geographical existence, sandalwood can be further grouped as East Indian sandalwood, Australian sandalwood, Hawaiian sandalwood and sandalwood of the Pacific islands. Apart from *S. album*, other species of *Santalum* are only of pedogeological interest as they are least exploitable for any kind of commercial purposes. *Fusanus spicatum* of Australia, *Eucarya accuminata* of

South Australia, East African species such as *Osyris tenuifolia*, *O. laceolata* and *Amyris balsamifera* L of West Indies are used as common adulterants of Indian sandalwood oil. The oil obtained from Indian sandalwood is valued high for its sweet and lasting bouquet. It has spicy, warm wood notes and has non-varying compositions with fixative properties. It is very rich in isomeric alcohols α- and β-santalol. Sandalwood oil is one of the oldest perfumery materials in the world. For decades, traditionally India has remained undisputedly dominant in sandalwood oil production for perfumeries and pharmaceutical use (Hansda 2009). Oil has a significant role in indigenous system of medicine and aromatherapy. It is used as antiseptic, anti-inflammatory, antiphlogistic, antispasmodic, astringent, cicatrising, carminative, diuretic, disinfectant, emollient, expectorant, hypotensive, memory booster, sedative and a prominent export material (Venkatesha 2011). Furthermore, it acts as a repellent against vermin and mosquitoes.

2.2.3 Indian Sandalwood in Vedic and Indus Valley Period

India possesses an age-old spectacular tradition of religious literature dating back to three and a half millennia. These epic monuments of human excellence are the narrations on the mysteries of the cosmos, the divine and mankind's relation with trees. The essence of Vedic Hymns was to praise and please the Gods with unique and divine offerings in return of requested favours. The Vedas depict that the Indian sandalwood was the most favoured item of the Gods. Tree worshipping was practised in India from the times of Harappan civilization. Vedic literature is said to be oldest penned document on plant life and it narrates that plant life began on earth before even the times of Devas (Gods). Vedas are the highest form of intelligence a human mind can achieve and the Vedic scholars had deep intuition to unravel the greatness and hidden values of plants.

The ancients of Sindhu valley were aware of the medicinal importance of plants and had put forward the concept of Ayurveda and wound healing since the Vedic period. It is mentioned in the epic Ramayana that Lord Lakshmana's wounds in the battle were healed with the help of 'Sanjeevani', an extract from a rare medicinal plant. In Vedic and Indus valley sandalwood was used in the form of artefact or carving deities, ritual bathing of deities, construction material for temples and for ceremonial utility (Kushalapa 1998; Fox 2000). The importance of Indian Sandalwood in Indian culture is cited in Vedic literature, referred in *Puranas* (ancient Hindu texts eulogizing various deities), exemplified in subhashitas (collection of eloquent sayings written by ancient Sanskrit poets), glorified in mythology (stories on supernatural beings and events) and mentioned in the ancient scriptures (sacred writings). Sanskrit manuscripts older than 4000 years refer to Indian sandalwood oil as the oldest known perfumery material. Indian sandalwood was used in sacrificial fires and is mentioned in Yajur Veda (Anonymous 1972). Likewise, melodious tunes loud hymns in 'Rig Veda' and delightful hymns of 'Sama Veda' stress upon the importance of Indian sandalwood. Sandalwood finds a fair share of mention even in the oldest of Indian literatures such as *Milinda Panha* (200 B.C.), *Patanjali Mahabhasaya* (100 B.C.), *Dhamma Pada*, *Anguttara Nikaya* (The teachings of Buddha) and *Vinaya Pitaka* (300–400 B.C.), etc. A Buddhist sage in *Milinda Panha* says that '*No flower's scent can waft against the wind, not sandalwood's, nor musk's, nor jasmine flowers, But the fragrance of the good goes against the wind in all directions the good man's name pervades*' while referring to righteousness. In a '*nibbana*' of *Milinda Panha* the sage again says '*Like sandalwood, it is hard to get, its fragrance is incomparable and it is praised by good men*'. Charaka and Susruta, the two pioneers of ayurvedic medicine gave an elaborate description of *S. album* and named it as '*Sweta Chandana*' (white sandalwood). The name of '*Sweta Chandana*' was continued for many ages and *S. album* was known by the common name of white sandalwood.

2.2.4 Indian Sandalwood in *Puranas* and Mythology

Data and documentation of the 'Sacred Plants of India' are based on its ethnomedicinal properties and modern-day research should emphasize on its importance. The traditional practices of Indians reflect the symbiotic association between humans and nature. The sacred trees of ancient period reflect the in-depth knowledge of advanced civilization that existed thousands of years ago and establish a unique principal portion of Indian ecological heritage. Trees were an inseparable part of Indian tradition since early ages. The traditional people of south India still believe that growing *S. album* in garden or backyard would ward off evil spirits and black magic. Some traditional households in south India are aware of the fact and commonly plant sandalwood in their home gardens for wealth and prosperity. This practice is gaining importance and is considered as a secret age-old tradition. Many households in Coimbatore grow *S. album* along with *Murraya koenigii* in their home gardens. It is also planted in the home gardens of Kerala localities in a similar fashion.

The age of Puranic literature is that of an intermediate position and depicted between Vedic and the period of classical literature (Pusalker 1963). There were about 18 major canonical *Puranas* written between these periods. Among them the most significant are *Agni Varta Purana*, *Brahma Vaivarta Purana*, *Bhagavata Purana*, *Kurma Purana*, *Matsya Purana*, *Skanda Purana*, *Shiva Dharma Purana*, *Vamana Purana* and *Vayu Purana*. The mention of Indian sandalwood can be seen many times in *Brahma Vaivarta Purana*. The Indian sandalwood is taken as the best example of goodness and fair practices. *Brahma Vaivarta Purana* mentions that the person owning/planting Indian sandalwood would be always blessed with wealth and glory. *Brahma Vaivarta Purana* signifies the importance of Indian sandalwood and refers that the very Goddess of Wealth 'the *Goddess Lakshmi*' herself resides in it and therefore the tree is believed to be auspicious and bringer of wealth. Furthermore, as per '*Kurma Purana*' the plants were created on earth through the dermal pores of Lord Brahma. The celebrated saint of Hindus 'Vyasa Rishi' writes about the many uses of Indian sandalwood mentioned in the '*Puranas*'. Likewise, the ancient Hindu culture glorifies the importance of many trees and associate those sacred trees with wisdom, knowledge, hidden secrets and consider them as sacred and divine. It can be said that the trees glorified in the ancient texts had either many medicinal values or were the important 'Keystone species'. According to *Vamana Purana* the worship of '*Lord Shiva*' with the beige-coloured paste of Indian sandalwood is recommended. '*Agni Varta Purana*' mentions about the plants to be used during religious ceremonies. These plants are important in performing 'Homma/Havana'—a Hindu ritualistic practice wherein offerings to the gods are made into a fire on special occasions. Now, it is a proven fact that these Homma/Havana purify the air or improve the quality of air. *Brahma Vaivarta Purana* exemplifies the significance of Indian sandalwood and mentions that Brahma created it through meditative contemplation.

In Indian Mythology the terms '*Vriksh* and *Vishnu*' means widely spread and omnipresent. Thus, it signifies that 'Lord Vishnu' is present in every form of life including plants. According to Indian mythology and as per legend, *Goddess Parvati* created a statue of a child using turmeric and sandalwood paste. The statue aftermath was blessed to life and is known as '*Lord Ganesha*' (God of intellect). An appreciation hymn (*Bhagavata Purana*) mentions that two holy brothers (*Lord Krishna* and *Lord Balarama*) while on their way to *Vrishni* Kingdom come across a dwarf named '*Kubja*' carrying sandalwood paste. Later, the dwarf lady gets blessed to be the beautiful woman after she gifts the paste to the two divine brothers. Origin of Indian sandalwood is even mentioned in the '*Panchatantra*' stories and was written that it was endemic to the Malaya mountains of southern India. A Poet in Ramayana while refers to river '*Tamara Parni*' and says that 'river islands covered with forests of sandalwood goes down the sea as a beloved maid to her lover'.

2.2.5 Indian Sandalwood in Various Religious Practices

'Brahma' the creator of Universe is regarded as the ultimate source and the saviour of Universe. Therefore, the ancient Indians worshipped plants and named a handful as sacred trees. Indian sandalwood is commonly used in the ritualistic practices of Hindus and is considered as divine. Apart from original religion of India it is also used in the sacred fires of Zoroastrian temples. In Zoroastrian culture it is believed that the use of Indian sandalwood in sacred fires soothes the troubles of all humanity. Likewise, Egyptians and Ceylonese use it for embalming mummies and medicinal purposes respectively. The practice of making incense sticks from Indian sandalwood dates back to 4000 years old (Adkoli 1977). The sandalwood incense sticks are a much sought-after commodity for prayers in Buddhist religion. It has been referred to as an item of toiletry in Buddhist Jataka stories of seventh century B.C. Early thinkers and herbal laureates believe that the scent of sandalwood incense promotes an atmosphere of peace, calms the mind and encourages spiritual awareness. The incense stick made out of Indian sandalwood is even used in Catholic Church and Jewish Synagogues. The major religion of India considered plants as dear to Gods and planting them became an act of worship. In Hindu epics and literatures, Kalpa-vriksha (*Adansonia digitata* L.), Chaitya-vriksha (*Ficus religiosa* L.) and Lakshmi-vriksha (*Santalum album* L.) are frequently mentioned which indicated the common practice of tree worship as an Indian tradition (Sandeep 2016). Indian sandalwood was also one of the most sought items of commerce from India that was exported to the Roman Empire. The quotes of Lord Buddha and Nobel Laureate Gurudev Rabindranath Tagore narrates its significance and importance. Lord Buddha describes the scent of sandalwood as the 'Fragrance of excellence'—'Sandalwood, tagara, lotus, jasmine —the fragrance of virtue is unrivalled by such kinds of perfume'. Likewise, our noble Laurette Gurudeva Rabindranath Tagore describes it as 'the Tree of Extraordinary qualities'—'The sandalwood tree as likely to prove, sweetest that can even conquer hate, love and perfumes the cruel axe that lays it low'. Significance of this species is underlined by the fact that sandalwood imparts fragrance even to the axe that cuts it. The concept of sacred trees was first conceived in ancient India and is believed that long ago sandalwood was found abundant in Coorg (Kodagu) (Fischer 1928). A famous district of Karnataka (Kodagu) is blessed with more than 600 sacred grooves and is commonly called as '*Devara Kadu*' in Kannada which spreads over an area of 10,000 acres. Traditional India is a country of sacred deities with many temples that conserve trees in its surrounding. This unique concept act as a symbol of sacredness and secular power. Nine important tree species are named after the nine planets of our solar system. In Sanskrit hymn they are eulogized as *Arkasam idam Adithygya* (*Calotropis gigantea* represented as Sun), *Palashagam Somaya* (*Butea monsperma* represented as Moon), *Khadiram Angarakaya* (*Acacia caechu* represented as Mars) *Apamargam Bhudhaya* (*Achyranwhus aspera* represented as Mercury), *Ashwatham Brihaspathaye* (*Ficus religiosa* represented as Jupiter), *Audumbarag Shukraya* (*Ficus glomerata* represented as Venus), *Shamigam Shaiaischaraya* (*Acacia ferruginea* represented as Saturn), *Rahuve Doorvaya* (*Cynodon dactylon* represented as Rahu) and *Kethuve Kushaya* (*Saccharum spontaneun* represented as Kethu). During '*Snapana Thirumanjanam*' (celestial bath), pre-dawn ritual '*Suprabatham*' and last seva '*Ekantha seva*' of *Lord Venkateshwara* huge quantities of sandalwood is required by Tirumala Tirupati Devasthanam (TTD). To satisfy its own overwhelming demand, TTD owns a sandalwood plantation spanning an area of 100 hectares near *Parveta Mandapam*, 3 km from the actual shrine of *Lord Venkateshwara* in Tirumala, Andhra Pradesh. Recently, Art of Living Foundation has also started plantations of Indian sandalwood in its premises under the aegis of

Shri Ravi Shankar Guruji. Likewise, Isha Foundation has started a unique initiative to plant and conserve sandalwood in India. The Royals frequently used sandalwood in funerals of their dead. Even to this day many royals practise the burning of their dead with sandalwood log materials. It has been customary in India to add a least possible of one piece of sandalwood to the funeral pile. All the religions across India worship trees in their own unique manner. Thus, all the religions of India promote the practice of plantation and advocate a healthy habit of conservation.

2.2.6 Indian Sandalwood of New Era

A renowned Poet of Karnataka, Kaviratna Kalidasa (fifth century A.D.) in his poem 'Raghuvamsa' refers to sandalwood at various instances. Sandalwood Trade, exchange and its origin are written in the texts of Periplus of the Erythrean Sea (written in the second century). Kautilya described the exchange and export of different classes and varieties of sandalwood in his writings of Arthasastra (200 B.C.). Sandalwood is frequently mentioned in the stories of Panchatantra and the significance of Indian sandalwood is reflected in Kavya—Mimamsa of Rajashekara (880–920 AD). A Tamil poet (Ilango Adigal) in his age-old writings mentions that Indian sandalwood actually originated from the Southern Mountains (Malaya Tenmalai Piranda Chandanam) of Peninsular India (Srinivasan et al. 1992). Amarkosha (Lexicon 3rd Century A.D.) glorifies the quality, usage and export of sandalwood to various European countries (Majumdar 1941). The Indian tradition worships trees and among them many are associated with the village or temple or the deity. TTD requires large quantities of sandalwood paste on a day-to-day basis. Interior walls and floors of the sanctum sanctorum of Lord Venkateshwara, small shrines and pooja rituals require a rich paste of sandalwood during 'Brahmotsavam' festival. Hindu God, Lord Venkateshwara is commonly known as 'Chandana Charchita Govinda' (The English narrative would be 'The Lord of the untainted form, The Lord who is beautiful and anointed by sandalwood paste, The Lord who has body tint equal to that of cloud') (Sandeep and Manohara 2019). Ethnomedicinal implications of sacred trees were written in the Vedas and Ayurvedic classics and were passed to the future generations by the practice of 'Gurukula Vidyabhyasam'. Though the country is enriched with impeccable traditional socio-religious values, hardly any progress is seen in ethnomedicinal repurposing and implications by modern Indian communities. Although the use of sandalwood dates back to the Vedic period, the original decline of natural stands of sandalwood started only from the period of Tipu Sultan.

2.2.7 Factors Leading to the Decline of Natural Populations of *S. Album* in India

Ancient Era gave utmost importance to nature and came up with unique ways of conservation. Whereas, the nature of new era is at the mercy of greed and sinister desires of mankind in the name of expansion, urbanization and development. The real decline in the sandalwood population started during 1792, when Tipu Sultan, de facto ruler of Mysore declared sandalwood as a royal tree. Tipu Sultan belonged to an Arab migrant tribe (Quraish) who used to frequent Turkey and Afghanistan for business, trade and for new war tactics. New age researchers and few historians review Tipu as a war monger who never hesitated to seek the help of Taliban and French to expel the one, who he hated. He applied skewed policies to get maximum benefit from sandalwood and various other trade (Lafont 2001). He reintroduced the concept of renaming of places and Mangalore was renamed as Jalalabad, Mysore as Nazarabad, Cannanore as Kusanabad, Dharwar as Quarshed-Sawad, Ooty as Faiz-Hissar, Dindigul as Khaliqabad, etc. However, many historians view him as a progressive king who introduced best global technologies to his Kingdom. Tipu introduced a unique system of stock-yard system for storing sandalwood known as 'Sandal Koti' in 1799, Koti meaning stockyard of sandalwood. Tipu Sultan was close associate of Afghans, French and Turks who generally favoured sandalwood. The Era of Tipu witnessed an extreme pinnacle of sandalwood export to Arabian countries. In the honour of his friendship, he sorted sandalwood in various classes and named it after Afghan, Arabic and Turkish terms. Likewise, in late 1898, the Government of Karnataka categorized 18 grades of sandalwood for the purpose of sale and auction. To establish a monopoly over trade, Tipu implemented a decree. This decree of 1792 established the monopoly of Indian sandalwood and led to the severe deterioration of sandalwood populations in natural stands. The Governments of Karnataka and Tamil Nadu later implemented a similar mandate to keep sandalwood under its control. The skewed Government policies and the Tipu decree was seen as a curse than a boon as it brought attracted more loss than benefit and several legal obligations were implemented. Earlier the full extraction and disposal of Indian sandalwood came under the control of Forest Department of Mysore state (Adkoli 1977). Overexploitation and illegal felling were a major cause for depletion of sandalwood gene pool in its natural stands (Naseer et al. 2012). It resulted in the loss of adaptive gene complexes (Gillieson et al. 2008). Forest fires, grazing, browsing; unconventional monoculture eucalyptus plantations, epidemic of spike disease (Mycoplasma like organism) and other biotic interferences (*Fusarium*, *Pythium*, *Rhizoctonia*, nematodes, etc.) have further contributed to the loss of this species. Its existence in natural stands is rare and tough to trace commercially exploitable sandalwood trees in its natural stands (Swaminathan et al. 1998). Few populations of higher girth classes can be seen in the protected boundaries of governmental and non-governmental organizations of south India.

2.2.8 Retrospective and Prospective Methods of Sandalwood Cultivation

Tribal culture of India has been a proud symbol of excellence in conserving natural resources of India. Although the implementation and usage of traditional knowledge in India saw some success in conservation a major contribution is yet to come from various governmental agencies to effective conservation of Indian sandalwood. The tradition and traditional knowledge need to be communicated and adapted by many societies across India for a meaningful conservation of this species. Large volumes of indigenous knowledge have never seen daylight and are cocooned in local tribes due to lack of proper publicity and awareness. Well-documented treatises of Charaka and Susruta are witness of advanced medical ayurvedic practices. Harsh government regulations have seen a tremendous fall in sandalwood populations. The traditional knowledge is gaining importance and playing a crucial role in addressing healthcare segments of the developing world. Traditional methods of sandalwood cultivation in India included the practice of dibbling and sowing seeds (Rai and Kulkarni 1986). The method of dibbling seeds was carried out just during the onset of rainy season and this method proved out to be the best during those times which led to formation of dense sandalwood forests in the semi-arid tracts of Mysuru. The tribes and early methods mostly used bamboo poles of about 1.5 m length with an internal diameter of 4–6 cm for sowing sandalwood (4–6) seeds in the bushes. The common traditional and new methods for the regeneration are (a) dibbling of sandalwood seeds in bushes (b) sowing of seeds on pits and mounds by trench mound method and (c) planting of nursery-grown seedlings (Sandeep and Manohara 2019). In the current scenario, the most prospective and lucrative method of sandalwood cultivation is the method of using nursery-grown seedlings. Generally, two types of seed beds are prepared for raising seedlings (sunken or raised). Raised beds are used in the regions with high rainfall. Healthy seedlings of about 30 cm height with dark brown bark of 6 months are best suited for planting. Approximately 2500 healthy seedlings can be produced from 1 kg seeds. After relaxation in constrictive government policies for sandalwood cultivation there is an enormous demand for its saplings in India.

Between 2003 and 2015 hundreds of plantations were conceived across India out of which many were unorganized and improperly managed. Due to inadequate consultation from nursery practitioners and non-scientific consultants many of the sandalwood plantations were mismanaged. Indian sandalwood can be grown along with agricultural and horticultural crops in different combinations (Joshi and Viswanath 2011). Studies have indicated that though *S. album* may grow well across different states in India but there is no uniform pattern in heartwood formation and oil content with the increasing age which may be attributed to several complex factors noted elsewhere. However, if the financial viability of sandalwood and increasing price trend of its oil in international market are considered then growing sandalwood tree is undoubtedly viable proposition that could boost the local economy (Viswanath 2010). Studies carried out in several locations of India revealed that heartwood formation and oil formation has been happening in other areas besides peninsular India (Joshi and Viswanath 2011; Sandeep et al. 2016; Babita et al. 2018). Currently there has been a sudden spurt in sandalwood cultivation across India and several successful stories are also a witness of adaptability of sandalwood. This preliminary study is expected to instil confidence among sandalwood growers in other parts of the country and encourage more farmers as well as private entrepreneurs to grow sandalwood by incorporating this tree in existing agroforestry practices.

2.3 Conclusion

Exhaustive exploitation has threatened the survival of *S. album* in its natural stands. The usage of sandalwood continues from the times of Puranas till the age of Artificial Intelligence (AI). In conservation of sandalwood the traditional knowledge has played a vital role and few households still practise planting it in their home gardens. Tree worship was part and parcel of ancient societies and tradition. It is still part of current tradition. The traditional knowledge and ancient findings have to be explored and followed for an effective conservation programme. For countries like India promotion of Sandalwood in agroforestry is also a good option as it not only helps in conserving this flagship species, but makes the traditional farming more productive, sustainable and profitable. It can act as a financial lifeline to the farmers. The Epics, Scriptures, Vedas, Puranas, Buddhist texts glorify the significance of *S. album*. The sacredness of *S. album* is witnessed in Indian mythology. Traditional belief unknowingly has paved the way for effective conservation of this flagship species but steps should be also taken by forest departments for long-term conservation. If Governments could bring in policy changes and liberalize existing rules and regulations regarding trade of sandalwood, in the long run it certainly makes pace shift changes and lures the attention of more productive farmers and encourages them to take up sandalwood cultivation.

Acknowledgements The authors are thankful to the Director, Institute of Wood Science and Technology, Bengaluru and to the Director, Kerala Forest Research Institute, Thrissur, Kerala. Authors are thankful to the National Medicinal Plants Board (NMPB), New Delhi.

Financial support from the Council of Scientific and Industrial Research (CSIR), Government of India through Senior Research Fellowship to Sandeep C (09/813(0003)/2015 EMR I) is gratefully acknowledged.

References

Adkoli NS (1977) Sandalwood in Karnataka. In: Retrospect and prospect: proceedings of the All-India sandal seminar. Bangalore, Karnataka, India, pp 86–94

Annapurna D, Rathore T S, Joshi G (2006) Modern nursery practices in the production of quality seedlings of Indian Sandalwood (*Santalum album* L.) Stage of host requirement and screening of primary host species. J Sust For 22(3/4):33–55

Anonymous (1972) Santalum. Wealth of India. Publication and information directorate, CSIR, New Delhi, India, pp 208–224

Babita M, Chakraborthy S, Sandhya MC, Viswanath S (2016) Sandalwood farming in India: problems and prospects. Ind J Trop Biod 26(1):1–12

Babita M, Sandeep C, Sushant A, Sruthi S, Viswanath S (2018) Assessment of heartwood and oil content of Santalum album Linn. in natural and naturalized populations across contrasting edapho-climatic conditions in India. Indian Fors 144(7):675–685

Brand JE, Fox JED, Pronk G, Cornwell C (2007) Comparison of oil concentration and oil quality from *S. spicatum* and *S. album* plantations, 8–25 years old, with those from mature *S. spicatum* natural stands. Aus Fors 70:235–241

Das SC, Tah J (2013) Effect of GA_3 on seed germination of sandal (*Santalum album* L.). Intl J Cur Sci 8:79–84

De Candolle (1857) Prodromus systematis naturalis regni vegetabilis

Doran J, Thomson L, Brophy J, Goldsack B, Bulai P, Fakaosi T, Mokoia T (2005) Variation in heartwood oil composition of young sandalwood trees in the South Pacific (*S. yasi*, *S. album* and F1 hybrids in Fiji, and *S. yasi* in Tonga and Niue). Sandalwood Research Newsletter, pp 3–4

Durairaj P, Kamaraj M (2013) Assessment and conservation strategies for *Santalum album* in Manmalai RF of Thuraiyur range at Tiruchirappalli district. Int J Hum Arts Med Sci 1(1):112

Fischer CEC (1928) The original home of *Santalum album* L. J Indian Bot Soc 7:12–13

Fox JE (2000) Sandalwood: the royal tree. Biologist (london) 47:31–34

Gairola S, Aggarwal PS, Ravikumar GS (2008) Status of production and marketing of sandalwood (*Santalum album* L.) In: Proceedings national seminar on conservation, improvement, cultivation and management of sandal. Institute of Wood Science and Technology, Bangalore

Ganeshaiah KN, Uma Shankar R, Vasudeva R (2007) Bioresource and empire building: what favoured the growth of Vijayanagara Empire? Cur Sci 93:140–146

Gillieson D, Page T, Silverman, J (2008). An inventory of wild sandalwood stocks in Vanuatu. Aus Cen Int Agri Res, Canbera, p 118

Hansda R (2009) The outlook for non wood forest products in Asia and the Pacific. Working Paper No. APFSOS II/WP/2009/18. Food and Agriculture Organisations Regional Office, Bangkok, p 89

Joshi SC, Viswanath S (2011) Prospects of *Santalum album L* (Sandalwood) cultivation in India. Sou Indian Hor 60:13–14

Kadambi K (1944) A working plan for the Devarayanadurga group of forests. Government Press, Bangalore

Kemp RH, Namkoong G and Wadsworth FH (1993) The nature of forest genetic resources. In: Conservation of genetic resources in tropical forest management Principles and concepts. FAO Forestry paper 107. Rome, p. 514

Krishnappa HP (1972) Sandal tree, a dollar earning parasite. Myforest 8:1–5

Kushalapa KA (1998) Trade liberalisation in sandalwood. In: Radomiljac AM, Ananthapadmanabha HS, Welbourn RM, Rao KS (eds) Sandal and its products. Proceedings of an international seminar organised by the institute of wood science and technology (ICFRE) and Karnataka state forest department, Bangalore, India 1997. Australian Centre for International Agricultural Research, Canberra, Australia, pp 24–26

Lafont JM (Trans) (2001) The memoires of Lieutenant Colonel Russel concerning Mysore: In the service Historique de l'Armee de Terre, Chateau de Vincennes, Paris. In: Habib I (ed) State and diplomacy under Tipu Sultan: documents and essays. Tulika Books, New Delhi, India pp 82–107

Majumdar GP (1941) The white sandal. Sci Cult 6:492–495

Naseer M, Shashank M, Pramod K, Ansari SA (2012) Genotyping of *Santalum album* L. accessions through cross-species transferability of SSR markers of *Santalum austrocaledonicum* and *Santalum insulare*. San Res News 27:1–5

Pearson RH and Brown HP (1932) Commercial timbers of India, vol 2, pp 862–868. Government of India, Central Publication Branch, Calcutta

Pusalkar, AD (1963) Studies in the epic and Puranas, (English), 2nd edition, Bombay, Bharatiya Vidya, Bhavan, Bhavan's Book University 36 (1955) 1st edition: ibid., Bhavan's Book University 36:225

Radomiljac AM, Shea SR, McKinell FH, McComb JA (1998) Potential for irrigated tropical forestry in north Western Australia. Aust for 61(2):70–75

Rai SN, Kulkarni HD (1986) Sandalwood plantations. In plantation crops., Opportunities and constraints. Oxford and IBH Publishing Co. 1:295–300

Rama Rao M (1908) Sandalwood at Sea Level. Indian for 34(151–153):406–407

Rama Rao M (1911) The host plants of the sandal tree. Indian For 2(4):159–207

Ramaswamy MN (1956) Essential oil reserves of Mysore. Indian for 82:127–129

Sandeep C, Manohara TN (2019) Sandalwood in India: historical and cultural significance of *Santalum album L.* as a basis for its conservation. NeBIO 10(4):235–241

Sandeep C, Raju G, Riajur R, Boruah RK, Viswanath S (2016) Variation in heartwood and oil content of *Santalum album* L. In Assam and Karnataka. J Biol 3 (1):79–87

Sandeep C, Kumar A, Rodrigues V, Viswanath S, Shukla AK, Sundaresan V (2020) Morpho-genetic divergence and population structure in Indian *Santalum album* L. Trees 1–17

Sen Sarma PK (1982) Sandalwood—its cultivation and utilization. In: Atal CK, Kapur BM (eds) Cultivation and utilization of aromatic plants. Regional Research Laboratory, CSIR, Jammu Tawi, pp 395–495

Srinivasan VV, Sivaramakrishnana VR, Rangaswamy CR, Anathapadmanabha HS, Shankaranarayana KH (1992) Sandal—(*Santalum album* Linn.). Institute of wood science and technology. Indian Council of Forestry Research and Education, Dehra Dun, India, p 233

Swaminathan MH, Hosmath BK, Mallesha, BB (1998) The status of sandalwood in India. In: Radomiljac AM, Ananthapadmanabha HS, Welbourn RM, Rao KS (eds) Sandal and its products. Australian Centre for International Agricultural Research, Canberra, pp 3–5

Troup RS (1921) Silviculture of Indian trees, vol 3. Clarendon Press, Oxford, pp 799–819

Venkatesan KR (1980) A fresh look at the management of sandal. In: Proceedings of 2nd forestry conference. FRI and Colleges, Dehradun, India, pp 1101–1108

Venkatesha Gowda VS (2011) Global emerging trends on sustainable production of natural sandalwood. In: Proceedings of the Art and Joy of Wood Conference, Bangalore, India

Viswanath S (2014) Sandalwood, an unexplored treasure. Inside Fact Natl Mon 1(2):10–13

Viswanath S, Dhanya B, Purushothaman S, Rathore TS (2009) Domestication of sandal (*Santalum album* L.) in India: constraints and prospects. Asia Pac Agro 34:9–12

Viswanath S, Dhanya B, Purushothaman S and Rathore TS (2010) Financial viability of Sandal (*Sandalum album L.*) agroforestry practices in southern India. Indian J AgroF 12(2):14–22

Sandalwood: Traditional Knowledge from Ayurveda

3

S. N. Venugopalan Nair,
Tabassum Ishrath Fathima,
and Shilpa Naveen

Abstract

Candana (*Santalum album*) is woven into the culture and traditions of India for over two thousand years. It is commonly known as Indian sandalwood which is the oldest and precious source of natural fragrance with immense medicinal, religious, and commercial significance. The economic importance of *Candana* and its essential oil has made it into the RET (Rare Endangered Threatened) category and it is one of the vulnerable plant species (ArunKumar et al. in The IUCN red list of threatened species, 2019). *Candana* is used in the traditional Indian system of medicine like Ayurveda, Siddha, and Unani, which is used in treatment and prevention of a wide range of ailments. This chapter aims to capture the essence of traditional knowledge on *Candana* as documented in Ayurvedic literature from 1500 BC to 1900 AD, with detailed citations and descriptions, translated from ancient *Acharyas* such as *Caraka, Susrutha,* and *Vagbhata* have been covered. Approximately 1011 references of *Candana* documented in 20 Ayurvedic classical texts, 30 synonyms in Sanskrit denoting its morphological characters, cultural and medical uses, 372 vernacular names found in 10 different languages are a part of this chapter. Vernacular names and Sanskrit synonyms of *Candana* along with their descriptive meaning illustrate the importance and popularity of *Candana* on a wider scale. The sheer depth of Indian Knowledge Systems is demonstrated with Sanskrit slokas and their approximate English meaning with the explanation of the various uses of *Candana* as well as simple and compound medicinal formulations. Highly informative data sources and a bibliography with supportive evidence authentic in nature have also been provided for further referencing.

Disclaimer The information provided by the authors in this chapter is for educational purposes only. It is based on the information available in authentic Ayurvedic texts written in Sanskrit and every care has been taken to provide English translation and interpretation of Sanskrit terms to the best of our knowledge. We make no representation of any information expressed or implied, regarding accuracy, adequacy, validity and reliability of the information given in this chapter. The health information is provided for general information and does not provide any kind of health advice.

S. N.V. Nair (✉) · T. I. Fathima · S. Naveen
Centre for Traditional Knowledge & Informatics,
The University of Trans-Disciplinary Health
Sciences and Technology (TDU),
Bengaluru-560064, India
e-mail: venu.gopal@tdu.edu.in

T. I. Fathima
e-mail: tabassum.if@tdu.edu.in

S. Naveen
e-mail: shilpa.naveen@tdu.edu.in

M. Gowda et al. (eds.), *The Sandalwood Genome*, Compendium of Plant Genomes,
https://doi.org/10.1007/978-3-030-93394-4_3

3.1 Introduction

The Sanskrit etymology of the word चन्दन is "चन्दयतिआह्लादयतीति" that which brings happiness is Candana. Similarly, it is related to the Moon "candrah", "shining, glowing" and the Latin *candere*, to shine or glow. It arrived in English via Late Greek, Medieval Latin, and Old French in the fourteenth or fifteenth century (Wikipedia contributors 2021).

Candana (sandalwood) is used to worship Lord Shiva, the supreme lord in Hinduism, and is believed that Goddess Lakshmi exists in the sandalwood tree making it very religious. The heartwood of the tree made into a paste is an integral part of all rituals and ceremonies, to make religious utensils, to adorn the deities, to calm the mind during meditation and prayers. In the temples, it is distributed to all devotees who apply it to their forehead, neck, or chest. Preparation of the paste (prepared by grinding the heartwood by hand on granite/stone slabs using water) is entrusted only to priests when used in temples and ceremonies as it is considered fit only to be done by the pure. In Tirupati, Sri Venkateshwara temple, after the religious tonsure, sandalwood paste is applied for its cooling reason and being able to protect the skin. Hinduism and Ayurveda consider Candana to bring one closer to the divine hence it is one of the holiest elements in Hindu and Vedic societies (Sandeep and Manohara 2019).

The ancient Egyptians imported sandalwood and used it in medicine, for embalming the dead, and in the ritual, burning to worship the gods (Kumar et al. 2012; Sultana and Rahman 2018).

There is evidence in trade of Indian botanicals with the Middle East and Europe around the fifth century BC during the Persian conquest of both India and Greece. Sandalwood was used in India from at least the fifth century BC, it was almost entirely confined to Buddhist and Hindu peoples. In the West it appears not to have been known until the beginning of the Christian era, the earliest Roman reference being in the famous *Periplus of the Erythræan sea* (Somadeva 1924), evidence pointing out on the existence of an export trade in Southeast Asian sandalwood along with other aromatic woods and resins have been imported into Egypt and the eastern Mediterranean through Arabian ports (Villier 2001). Some of the ornamental and fragrant wood like ebony, teakwood, sandalwood, and other aromatic wood were traded (Bag 2016).

3.1.1 History of CANDANA in the Vedic and Indus Valley Period

Candana is associated with the Indian culture and traditions, from the times of Harappan civilization with rituals of tree worshiping was practiced in India. The Ayurvedic medicinal importance of Candana was known since Vedic period and Indus valley civilization, its use in woodworking artifacts or carving deities, ritual bathing of Hindu deities, temple construction material, and ceremonial causes (Sandeep and Manohara 2019). Its importance in Indian culture is referred to in Vedic literature, *Puranas* (ancient Hindu texts eulogizing various deities), *Subhashitas* (collection of eloquent sayings written by ancient Sanskrit poets), mythology (stories on supernatural beings and events), and in the ancient scriptures (sacred writings). The Sanskrit manuscripts that are more than 4000 years old cite sandalwood oil as the oldest known perfumery material. Literature evidence shows the use of sandalwood in sacrificial fires since Yajur Veda (Sandeep and Manohara 2019), while the treaties of loud hymns in Rig Veda and melodious hymns of Sama Veda give indirect evidence of Indian sandalwood. It is cited even in the oldest of Indian literature such as MilindaPanha (200 B.C.), Patanjali Mahabhasaya (100 B.C.), Dhamma Pada, AnguttaraNikaya (The teachings of Buddha), and Vinaya Pitaka (300–400 B.C.). In MilindaPanha (Buddhist doctrine), Buddhist sage while referring to the path of righteousness mentions that "No flower's scent can waft against the wind, not sandalwood's, nor musk's, nor jasmine flowers. But the fragrance of the good goes against the wind in all directions the good man's name pervades." While describing "nibbana" in MilindaPanha, the sage also mentions that "Like sandalwood, it is hard to get, its fragrance is incomparable and it is praised by good men."

3.1.2 Historical Data from Ayurvedic Works

The classical texts of Ayurveda have systematically documented its properties and usage of Candana over a period of the last 3000 years. We have systematically screened the following texts of Ayurveda as given in Table 3.1, which highlights the depth of traditional knowledge as documented in classical text during their corresponding period.

An approximate 1011 references on Candana are found documented in 20 Ayurvedic texts while writing this report (Nair et al. 2020).

3.1.3 Recent History

Candana oil has a great demand both in the international and national market, because of its economic value, and over-exploitation has subsequently been categorized as vulnerable in IUCN Red List. (Arunkumar et al. 2016). India has been the traditional leader of sandalwood oil production for perfumery and pharmaceuticals and it is also one of the second most expensive wood in the world (Kumar et al. 2012). Candana had been declared as the "royal tree" by the

Table 3.1 Classical Ayurvedic text selected for plant Candana

	Text name	Chronology	Author	Number of references on plant name Candana
1	Caraka Samhita	1500 BCE-200 CE	Agnivesa, redacted by Caraka,	160 citations on clinical use
2	Susrutasamhita	1500 BCE-500 CE	Susruta, Nagarjuna	108 citations on clinical use
3	AstangaSangraha	500 CE	Vagbhata	223 citations on clinical use
4	AstangaHrdayam	600 CE	Vagbhata	93 citation on clinical use
5	Astanga Nighantu	800 CE	Vagbhata	Listed as Candana
6	Paryayaratnamala	900 CE	Madhava	Listed 4 synonyms
7	Dhanvantari Nighantu	1000 CE	Unknown	Listed 14 synonyms and properties
8	Cakradatta	1075CE	Cakrapanidatta	133 tations on clinical use
9	Sarngadharasamhita	1300 CE	Sarngadhara	42 citations on clinical use
10	Nighantu Sesa	1200 CE	Hemachandra	Listed 13 synonyms and properties
11	Siddhamantra	1210 CE-1247CE	Kesava	Listed synonyms and properties)
12	Hrdayadipaka Nighantu	1260 CE-1271 CE	Bopadeva	listed synonyms and properties
13	Madanapala Nighantu	1374 CE	Madanapala	Listed 7 synonyms and properties
14	Bhavaprakasha	1550 CE	Bhavamisra	225
15	Bhavaprakasha Nighantu	1550 CE	Bhavamisra	listed 7 synonyms and properties
16	Raja Nighantu	1700 CE	Naraharipanda	Listed 13 synonyms and properties
17	Saligrama Nighantu	1896 CE	Saligramavaisya	Listed many synonyms and properties compiled from other texts
18	Siddhabhesajamanimala	1896 CE	Krshnaramabhatta	Listed one main name as Candana

Sultan of Mysore in 1792, by Tipu Sultan the ruler of the kingdom of Mysore, and further took over sandalwood trade of the state on a monopoly basis around 1792, the practice was continued by the later Maharajas of Mysore (Nalwadi Krishnaraja Wodeyar) and subsequently by the Karnataka Government until recently. Karnataka and Tamil Nadu adopted the monopoly and kept sandalwood under government control and many legal obligations were imposed on sandalwood cultivation (Pallavi 2015).

3.2 Nomenclature Studies

Ayurvedic Samhitas follow polynomial nomenclature methods. This nomenclature system is useful in identifying the plant species based on their morphological, cultural, and clinical importance as authors of various lexicons or "*nighantus*" recorded them in different time periods. The purpose and reason for this method is given in *Dhanvantarinighantu* during the tenth century AD.

वहून्यथ प्राकृतसंस्कृतानि नामानि विज्ञाय वहूंश्च पृष्ट्वा ।
दृष्ट्वा च संस्पृश्य च जातिलिङ्गैर्विद्यात् भिषग्भेषजमादरेण ॥

(Dhanvantari nighantu)

Plant names which have an origin in oral traditions, local dialects, languages, and written documents represent the rich knowledge tradition of a given region. It also explains various taxonomic classifications, habit, habitat, morphological explanations properties, and usage of plants in a broad sense but adequate for a medical practitioner. An analysis of the used nomenclature of plants gathered from various sources needs to be confirmed by questioning, reasoning, and observing. One has to go further by feeling them by touching the plant parts to confirm their identity. This will also help us to understand the purpose of ascribing different names to them. Ayurvedic lexicons also use different names with minor gender variations denoting the features of the plant or sometimes it is ascribed to different species itself, for example "Abhaya" and "Abhayaa" are two different species as documented in the Sanskrit language.

3.2.1 Sanskrit Synonyms of Candana

The polynomial nomenclature system used in Ayurveda has documented around 30 names over a period of time in Sanskrit denoting its morphological characters, cultural and medical uses.

1. Bhadrasri-pleasant, auspicious (Karp, 11—BPN, Cand, 6/7—RJN)
2. Bhadrasriyam-pleasant, auspicious (Cand, 3, 1—DHN, Karp, 8—MPN)
3. Bhogivallabham-snakes are found often around the tree (Cand, 6/7—RJN)
4. **Candanam/Candana**-pleasant to see (Karp, 11—BPN, Cand 3, 1—DHN, Karp, 8—MPN, Cand, 6/7—RJN)
5. Candrayutih-cooling and pleasant to see as the moon (Karp, 11—BPN)
6. Gandhadhya-rich in odor, very fragrant (Cand, 6/7—RJN)
7. Gandharaja-best among all aromatic substances (Cand, 6/7—RJN)
8. Gandhasara-aroma is situated in the heartwood (Cand, 6/7—RJN)
9. Gandhasarah-aroma is situated in the heartwood (Karp, 11—BPN)
10. Gandhasarakam-aroma is situated in the heartwood (Karp, 8—MPN)
11. Gandhasaram-aroma is situated in the heartwood (Cand, 3, 1—DHN)
12. Gosirsa-(Cand, 6/7—RJN)
13. Gosirsam-(Cand, 3, 1—DHN, Karp, 8—MPN)
14. Maharham (Cand, 3, 1—DHN, Karp, 8—MPN)

15. Malayajah-grows in Malaya region (Kar, 11 —BPN)
16. Malayajam-grows in Malaya region (Cand, 3, 1—DHN, Karp, 8—MPN, Cand, 6/7—RJN)
17. Malayodbhavam-Origin in the Malaya / hilly region (Cand, 6/7—RJN)
18. Mangalya-auspicious, prosperous (Cand, 6/7—RJN)
19. Pavanam-pure, used in means of purification (Cand, 6/7—RJN)
20. Sarpavasa-serpents are often found near the tree (Cand, 6/7—RJN)
21. Sisira-it is coolant
22. Sitaganda-having cool fragrance (Cand, 6/7—RJN)
23. Sitala-cooling in nature (Cand, 6/7—RJN)
24. Srikhandam-that which is beautiful, good (Karp, 11—BPN, Cand, 3, 1—DHN, Cand, 6/7—RJN)
25. Sugandha-aromatic, fragrant (Cand, 6/7—RJN)
26. Svetacandanam (Cand, 3, 1—DHN, Karp, 8 —MPN, Cand, 6/7—RJN)
27. Tailaparnikah-leaves are similar to those of the sesame plant (Karp, 11—BPN)
28. Tilaparna-leaves are similar to those of the sesame plant (Cand, 6/7—RJN)
29. Tilaparnakam-leaves are similar to those of the sesame plant (Cand, 3, 1—DHN)
30. Tilaparnam-leaves are similar to those of the sesame plant (Karp, 8—MPN) [RJN-Raja nighantu, MPN-Madanapalanighantu, BPN-Bhavaprakashanighantu, DHN-Dhanvantarinighantu].

3.2.2 Vernacular Names of Candana-

There are a total of **372** Vernacular names found in **10** languages. The total number of names recorded in each language and the widely used names are listed below.

English (1)—Indian sandalwood.

Hindi (11 names)—Candal, Candan, Candoi, Safed candan.

Kannada (33 names)—Agarugandha, Bhadra shree, Bhadrasri, Candana.

Malayalam (19 names)—Candana.

Marathi (3 names)—Candan, Gandhachakoda, Safed candan.

Sanskrit (58 names)—Candana, Gandhasara, Bhadrasree, Pitachandana, Shrigandha, Svetacandana.

Tamil (187 names)—Cantana, Srigandam, Sandana.

Telugu (51 names)—Bhadrasri, Chandanam, Gandhamu, Haricandanamu, Shreegandhamu, Sirigandamu, Tellacandanamu.

Tibetan (4 names)—Dkar-po tsan-dan, Tsandanadkar po, Tsa-ndan, Tsan dan.

Urdu (5 names)—Buradasafaid, Burada sandal safaid.

[Anonymous, 2017; Anonymous, 2016].

3.3 Ayurvedic Profile

3.3.1 Classification of Candana

Caraka Samhita (1500 BCE–200 CE): Candana has been classified among the analgesic group of herbs *(Angamardaprasamana)*, the ones that relieve burning sensation *(dahaprasamanagana)*, group of drugs that are useful in skin disorders (*kandughnagana*), drugs that are predominantly bitter in taste (*tiktaskandadravya*), thirst-quenching group of drugs, (*trsnanigrahanagana*), group of herbs that enhance complexion (*varnyagana*), and herbs that are anti-toxic (*vishaghnagana*).

Susruta Samhita: (1500 BCE–500 CE): In Susruta Samhita, Candana has been grouped under medicinal drugs starting with *Salasaradigana*. This group helps to cure skin disorders, absorb the deranged fat and *kapha* and proves beneficial in morbid discharges from the urethra (Meha) chlorosis or jaundice (*pandu, kamala*).

Candana is also grouped under Patoladigana which is a febrifuge and anti-toxic, and its therapeutic action consists of destroying the action of deranged *pitta* and *kapha*. It restores the natural relish of the patient for food, mitigates vomiting, and proves beneficial in ulcers, itching, and erythematous eruptions.

Sarivadi group also contains Candana. This group of drugs alleviates thirst and proves curative in the case of hemoptysis. Its therapeutic virtue consists in curbing an attack of bilious (*pittaja*) fever and in specifically alleviating the burning sensation of the body.

*Guduchyadi*gana also contains Candana. Drugs in this group act as good appetisers, general febrifuge, and successfully combat symptoms such as nausea, want of relish for food, vomiting, thirst, and pacifies the burning sensation of the body.

Candana is grouped under Pitta samsamana group of drugs that pacifies *pitta dosha.*

considered to be an evergreen tree. The Sukvadi variety appears little dried always (Sastry 2005).

The botanical species correlated to the above classification are Pterocaupussantalinus(raktachandan), *Adenantherapavonia*(kuchandana), *Caesalpinia sappan*(pattanga, kaliyaka).

3.3.3 Cardinal Signs of Best Candana

श्रेष्ठं कोटरकर्परोपकलितं सुग्रन्थि सङ्गौरवं छेदे रक्तमयं तथा च विमलं पीतञ्च यद्घर्षणे |
स्वादे तिक्तकटुः सुगन्धबहुलं शीतं यदल्पं गुणे क्षीणञ्चार्धगुणान्वितं तु कथितं तच्चन्दनं मध्यमम् ||१|| (***Raja nighantu***)

Raja Nighantu, (Indradev, 2006). The heartwood taken from the nodular part of the Candana tree, which is heavy, red, and on rubbing yields a clear, white-yellowish paste, is considered the best Candana. It is pungent and bitter in taste, cooling, and very fragrant. The Candana which is slightly less in the above-said properties or even half of the aforesaid properties is considered as a medium.

3.3.2 Varieties of Candana According to Each Nighantu/Lexicons in Ayurveda

Dhanvantari Nighantu (1000 CE)-Five varieties-Candana, Raktacandana, Kucandana, Kaliyaka, and Barbarika.
Sodhala Nighantu Eight varieties-Gosirsa, Sripriya, Sambhrana, Kakubha, Barbarika, Raktacandana, Patanga, and Haricandana.
Kaiyadeva Nighantu Six types-Candana, Raktacandana, Pita candana, Kairatacandana, Barbara candana, and Hari candana.
Barbarika is again of three types viz. white (sveta), black (krsna), and red (rakta).
Three types based on the color-Red (Rakta), yellow (pita), and white (pandu).
Bhava prakasa (1550 CE)-Four varieties-Candana, Raktacandana, Kaliyaka, and Kucandana.
Raja Nighantu (1700 CE)—Six varieties-Candana, Raktacandana, Patanga, Kaliyaka, Barbaraka, and Haricandana.

Sveta candana is again of two types viz. Vetta and Sukvadi.

The Vetta variety described by Raja Narahari is found mainly in the Malaya mountains. This is

3.3.4 Major Pharmacological Properties of Candana

Taste (rasa)—bitter (tikta), sweet (madhura).

Biophysical properties (guna)—light to digest (laghu), dry (rooksha).

Potency (veerya)—Cold in nature (sheeta).

Post digestive effect (vipaka)—Undergoes taste conversion into pungency after digestion (katu).

Ayurveda explains any biological system based on its systemic understanding. "*Vata hara*" denotes, its ability to decrease the rate of all neurological activities due to its associated biophysical (e.g., *Guna*) and biochemical properties (e.g., *Rasa panchakas*). "*Pitta hara*" indicates its ability to decrease the rate of all metabolic activities due to its associated biophysical (e.g., *Guna*) and biochemical properties (e.g., *Rasa panchakas*). "*Kapha hara*" denotes its ability to decrease the rate of all anabolic activities due to its associated biophysical (e.g., *Guna*) and biochemical properties (e.g., *Rasa panchakas*). The outer layer of Chart 3.1 shows the end result of such systemic function in a given patient or a clinical condition.

Actions (karma)-eliminates body odor (*dourgandha hara*), pacifies burning sensation (*dahaprasamana*), enhances complexion (*varnya*), helpful in skin disorders (*tvakdosha hara*), promotes memory (*medhya*), pacifies morbid thirst (*trsnanigrahat*), pacifies vitiation of pitta dosha (*pitta samaka*), cordial to the heart (*hrdya*), purifies blood (*raktasodhana*), alleviates hemorrhagic disorders (*raktapittasamaka*), kaphanissaraka, diuretic (*mutra janana*), helps in sweat production (*svedajanana*), useful in skin disorders (*kushtaghna*), cures fever (*jvaraghna*), antidote (*vishaghna*) (Sharma 2006).

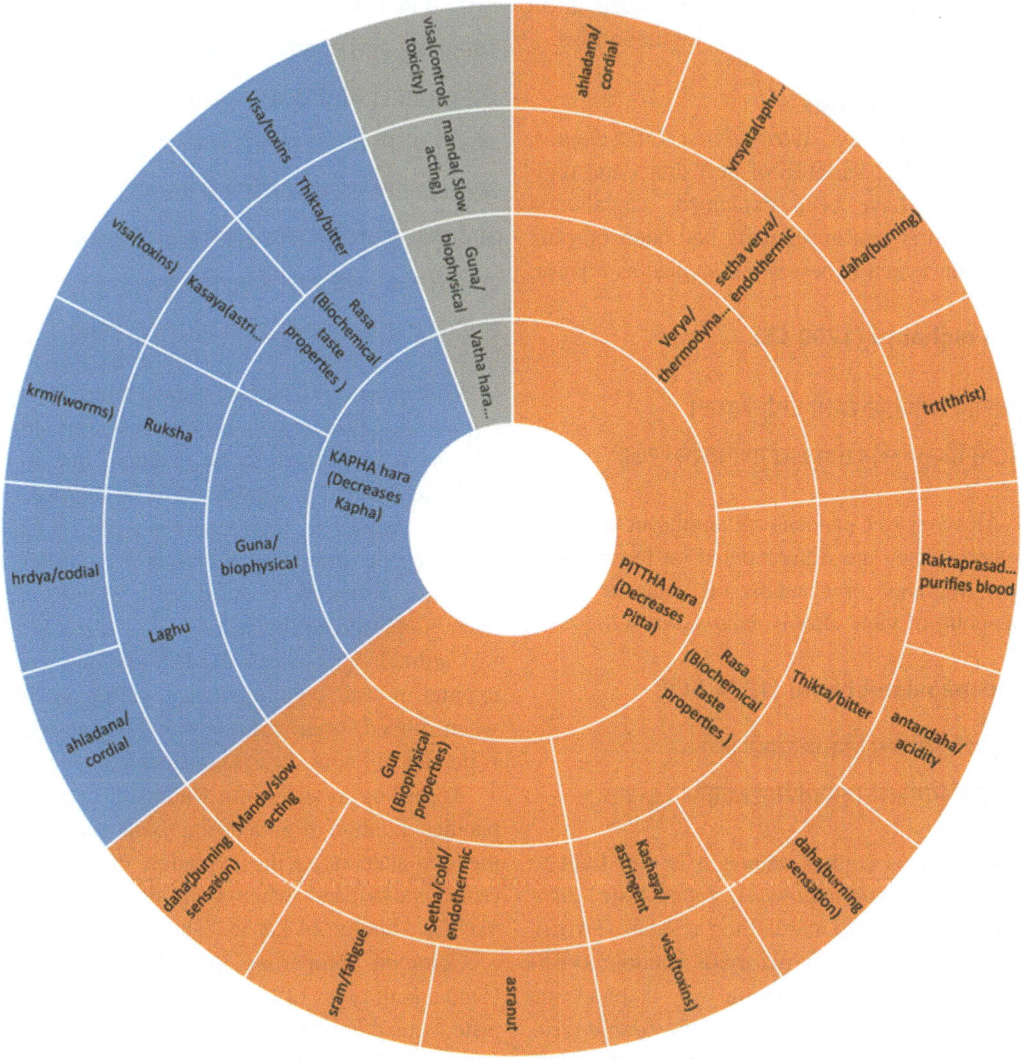

Chart 3.1 Ayurvedic understanding of the Systemic functions of Candana and its relationships with properties and functions

Usage according to seasons

Spring season-Juices fermented drinks and beverages of grape juice; *sharambu* (juice of sandalwood) is advised. One should smear his body with saffron, camphor, sandalwood, and *agaru* (*Aquilaria agallocha*). It reduces itching and body odor. This was the ancient way of perfume.
Summer-During nights after smearing the body with sandalwood paste one should sleep in an open airy roof which is cooled by moon rays.

Properties and uses

In *Caraka Samhita, sutra sthana 25th chapter*, among the 152 varieties of best drugs and regimens that have been explained, sandalwood (Candana) is best in relieving bad odor, burning sensation which is best when applied as a paste.

Raja nighantu (1700 CE)

चन्दनानि समानानि रसतो वीर्यतस्तथा ।
भिद्यन्ते किन्तु गन्धेन तत्राऽऽद्यं गुणवत्तरम् ॥२७॥

All types of Candanas are similar in taste and potency. They are differentiated on the basis of their fragrance or Candana oil contents. From this point of view, *Malayajacandana* is the best.

Madanapalanighantu (1374 BCE)

चन्दनं शीतलं रुक्षं तिक्तमाह्लादनं लघु ।
हृद्यं वर्ण्यं विषश्लेष्मतृष्णापित्तास्रदाहजित् ॥१॥

Candana is cold in nature, it is dry, light to digest (*sitaruksalaghuguna*), bitter taste (*tikta rasa*), *ahladjanana*, cordial to the heart (*hrdya*), enhances complexion (*varnya*), antidote (vishaghna), alleviates vitiation of kaphadosha (*kaphahara*), pacifies morbid thirst (*trsnanigrahana*), cures hemorrhagic disorders (*raktapittahara*), and pacifies burning sensation (*dahaprasamana*) (Sastry, 2010a, b).

Kaiyadevanighantu

श्रेष्ठं मलयसम्भूतं यवनेषु च मध्यमम् ॥१२६२॥
बर्बरेष्वधमं प्रोक्तं पाञ्चालयवनादिषु ।
सर्वाण्येतानि तुल्यानि रसतो वीर्यतस्तथा ॥१२६३॥
गन्धेन तु विशेषोऽस्ति पूर्वं श्रेष्ठतमं गुणैः ।
स्वादने कटु शीतं च घर्षणे पीतमेव च ॥१२६४॥
छेदने रक्तवर्णञ्च श्रेष्ठं चन्दनमुच्यते ।
अत्रापि चोत्तमं विद्याद् ग्रन्थिखर्परकोटरम् ॥१२६५॥
आर्द्रच्छिन्नं पित्तहरं शुष्कच्छिन्नं च वातनुत् ।
आर्द्रशुष्कं तु यच्छिन्नं तच्च श्लेष्महरं परम् ॥१२६६॥
सतिक्तं चन्दनं सर्वं सूक्ष्मं स्वादु हिमं गुरु ।
तृष्णादाहास्रपित्तघ्नं क्लमशोषविनाशम् ॥१२६७॥

With respect to place-Candana from Malaya region is considered to be the best, one from Yavana is considered to be medium, one from Barbara, Panchal is considered to be inferior.

With respect to fragrance-All types of Candana are similar in taste and potency but differ in fragrance. From this point of view, Malaya candana is considered the best.

Cardinal signs of best candana-the one which is bitter in taste, cold in potency, yellow in color when rubbed, when cut it is red. The one which is devoid of knots, etc., is considered as the best.

According to state/form-when wet or fresh it pacifies vitiation of pitta dosha (*pitta samaka*), in the dry state it pacifies vitiation of vatadosha (*vatasamaka*) and when both dry and wet it pacifies vitiation of kaphadosha (*kaphasamaka*).

Common properties of all varieties of Candana-all types of Candana are sweet-bitter in taste (madhura-tikta), cold in potency (sitaveerya), heavy to digest (guru), sookshma.

They help to cure morbid thirst (trsna), burning sensation (daha), hemorrhagic disorders (rak-tapitta), klama and emaciation (sosha) (Sharma, 2013).

Dhanvantarinighantu (1000 CE)

श्रीखण्डं शीतलं स्वादु तिक्तं पित्तविनाशनम् |
रक्तप्रसादनं वृष्यमन्तर्दाहापहारकम् ||२||
पित्तास्रविषतृड्दाहकृमिघ्नं गुरु रूक्षणम् |
सर्वं सतिक्तमधुरं चन्दनं शिशिरं परम् ||३||

Candana is sweet and bitter in taste, cold in potency, pacifies vitiation of pitta dosha, is blood purifying, aphrodisiac and useful in an internal burning sensation.

All kinds of candana are sweet and bitter in taste, cold in potency, pacified pitta and rakta-dosha, bad effects of poisons, thirst, burning sensation, and worms. These are heavy to digest, dry, and cause refrigerant action (Sharma, 2008).

Bhavaprakasanighantu (500 CE)

स्वादे तिक्तं कषे पीतं छेदे रक्तं तनौ सितम् |
ग्रन्थिकोटरसंयुक्तं चन्दनं श्रेष्ठमुच्यते ||१२||
चन्दनं शीतलं रूक्षं तिक्तमाह्लादनं लघु |
श्रमशोषविषश्लेष्मतृष्णापित्तास्रदाहनुत् ||१३||

Properties of best Candana-The heartwood taken from the nodular part of the Candana tree, which is heavy, red, and on rubbing yields a clear, white-yellowish paste, is considered the best Candana. Candana is light in digestion, dry, cold in potency, pacifies vitiation of pitta and kaphadosha, cooling in action, cures debility, thirst, poisoning, burning sensation, and diseases caused due to vitiation of blood (Chunekar, 2010).

3.4 Major Therapeutic Uses

3.4.1 Burning Urination with Blood

In case of burning micturition, Candana, cooked with rice and taken along with cool milk is beneficial (Srikantha, 2000; Sharma, 2007).

Candana paste taken internally with sugar candy and uncooked rice-washed water is useful in urination associated with bleeding.

3.4.2 Leucorrhoea

Candana is given with milk, ghee, sugar, and honey (Indradev, 2005).

3.4.3 Headache

Paste of Candana is a specific remedy (Sharma 2007a, b).

3.4.4 Burning Sensation

Paste of Candana is used (Sharma 2007a, b).

3.4.5 Skin Disorders

Paste of Candana is used (Sharma 2007a, b).

3.4.6 Hemorrhagic Disorders

Candana powder should be taken along with *tandulodaka* (rice-washed water) (Cs.Ci.4) (Sharma, 2014).

श्रृतशीतपयोऽन्नाशीचन्दनंतण्डुलाम्बुना
पिबेत्सशर्करंश्रेष्ठमुष्णवातेसशोणिते।। (चक्तदत्त, ३३-१४, भावप्रकाशमूत्ररोगधिकार३६-३९)

3.4.7 Bleeding Piles

Ginger and candana should be consumed (Cs. Ci.14/185) (Sharma, 2014).

3.4.8 Vomiting

Powder of candana should be taken along with gooseberry juice (Cs.Ci.20/33) (Sharma, 2014).

3.4.9 Bleeding Diarrhea

Fine powder of candana, honey, and rice-washed water should be consumed. It alleviates burning sensation, thirst, diabetes, etc. (Cs.Ci.19/86) (Sharma, 2014).

3.4.10 Menorrhagia

A decoction of svetacandana (white variety) or raktacandana (red variety) should be taken (Ss. Sha.2–15) (Sharma, 2000 & 2005).

3.4.11 Urticaria

Juice of guduci (*Tinospora cordifolia*) should be taken along with candana powder (***Vaidya manorama***).

3.4.12 Amrtacandana (Cd. 1/206)

Guduci (*Tinospora cordifolia*), Candana, Sunthi (*Zingiber officinale*) all these drugs are cleaned and chopped into pieces. oil 1/2 cup of material (1 cup = 150 ml) in 4 cups of water and reduce to 1 cup. Filter it and use. This amrtacandanakvatha is tridoshahara (alleviates all the three doshas) and is used in tritiyakajwara (a kind of visamajvara (malarial fever)) this is administered internally in a dose of 40–60 ml thrice a day, on empty stomach (Sharma, 2007).

3.4.13 Candana 12 (Ah. Ci. 4/3)

One part of candanakalka (paste) is boiled with 4 parts of ghee and 16 parts of the dravadravya (liquid). These are boiled till the liquid added has evaporated and the paste does not stick to the bottom of the vessel. It is administered internally in case of svasa (dyspnoea) (Srikantha, 2014).

3.4.14 Candana 25 (as. Ci. 11/24)

Well-dried materials of Candana, Sita, are crushed and made into a fine powder and sieved. The curna is pittahara (alleviates pitta) and is good in case of rakta atisara (diarrhea along with blood). It is administered internally in a dose of 5–10 gms thrice daily along with honey and sugar (Srikantha, 2012).

3.4.15 Candana (Cs. Ci. 3/261)

Candana boiled in water is called as Candana jala which is recommended in daha jvara (fever with burning sensation) (Indradev et al., 2003; Indradev, 2005; Sharma, 2014).

[Cs.Ci-Carakasamhita, cikitsasthana; Ss.Sha-Susrutasamhita, shaareerasthana;Cd-Cakradatta; Ah.Ci-Astangahrdaya, cikitsasthana; As.Ci-Astangasangraha, cikitsasthana] (Indradev, 2006).

3.5 Compound Formulations

There are around 110 compound formulations in Ayurveda (from selected texts) in which Candana is a major ingredient. The important compound formulations produced by the pharmaceutical companies are (1) Candanadivati (3) Candanaasava (4) Candanaditaila (5) Candanadivarti (6) Kunkumaditailam (Dasji, 2006).

References

Anonymous (2016) FRLHT's ENVIS centre on medicinal plants, FRLHT and MoEFCC. Government of India. envis.frlht.org / frlhtenvis.nic.in

Anonymous (2017) NMPB, Ministry of AYUSH. Government of India and FRLHT, Bangalore. http://www.medicinalplants.in/

Arunkumar AN, Joshi G, Rao MS, Rathore TS, Ramakantha V (2016) The population decline of Indian sandalwood and people's role in conservation—an analysis. In: Nautiyal S, Schaldach R, Raju K, Kaechele H, Pritchard B, Rao K (eds) Climate change challenge (3C) and social-economic-ecological interface-building. Environmental science and engineering. Springer, International Publishing, Switzerland, pp 377–387. https://14.139.155.5:2465/10.1007/978-3-319-31014-5_22

Arunkumar AN, Dhyani, A, Joshi G (2019) Santalum album. The IUCN red list of threatened species 2019: e.T31852A2807668. https://doi.org/10.2305/IUCN.UK.2019, 1.RLTS.T31852A2807668.en

Bag AK (editorial) (2016)Trade-links, commerce and exchanges of scientific ideas in antiquity (editorial). J Hist Sci 51(4):563–570

Chunekar KC (2010) Bhavaprakasa Nighantu, V Ayurveda Series, Chaukhambha Bharati Academy, Varanasi, Karpuradi varga. Sloka No 12–13:178

Dasji Govinda Shri (2006) Bhaisajyaratnavali, The Chaukhambha Sanskrit Bhawan series 67, Chaukhambha Sanskrit Bhawan, Varanasi, Vol III, Chapter 60, Sloka 115–120, 188

Indradev T, Nighantu R. Krishnadas (2006) Ayurveda series 5, vol 27. Chowkhambha Krishnadas Academy, Varanasi, Candanadi Varga, Sloka, p 400

Indradev T. Gadanigraha of Sri Vaidya Sodhala (2005) The Kashi Sanskrit series 182, Chaukhambha Sanskrit Sansthan, Varanasi, Vol 3, Kaumaratantre pradradhikara 1, Sloka 42, pp 453

Kumar A, Joshi G, Ram H (2012) Sandalwood: history, uses, present status and the future. Curr Sci 103 (12):1408–1416. http://www.jstor.org/stable/24089347. Accessed 24 Feb 2021

Murthy Srikantha KR (2000) Bhavaprakasa of Bhavamisra, Krishnadas Ayurveda series 45, Chowkhamba Krishnadas Academy, Reprint ed, Varanasi

Murthy Srikantha KR (2012) Astanga Sangraha of Vagbhata, Jaikrishnadas Ayurveda series 79, Chaukhambha Orientalia, Varanasi, vol II, Chapter 11, Sloka 24, 405, reprint edition

Murthy Srikantha KR (2014) Astanga Hrdayam, Krishnadas Ayurveda series vol 27, Chowkhamba Krishnadas Academy, reprint ed, Varanasi

Pallavi A (2015) Forests, return of scented wood. Down to earth, Feb 25th. https://www.downtoearth.org.in/coverage/forests/return-of-scented-wood-48569. Accessed 5 March 2021

Sandeep C, Manohara TN (2019). Sandalwood in India: historical and cultural significance of Santalum album L. as a basis for its conservation. NeBIO 10(4):235–241

Sastry JLN (2010a) Madanapala Nighantu, Jaikrishnadas Ayurveda series 114. Chaukhambha Orientalia, Varanasi, pp 368–370

Sastry JLN (2010b) Madanapala Nighantu, Jaikrishnadas Ayurveda series 207, vol 9. Chaukhambha Orientalia, Varanasi, Karpuradi varga, Sloka, pp 368

Sharma PV (2000) Suruta Samhita, Haridas Ayurveda series 9. ChaukhambhaVisvabharathi, Varanasi II:128–129

Sharma PV (2006) DravyaGunaVignana, V. Ayurveda series 3. Chaukambha Bharati Academy, Varanasi 2:715–717

Sharma PV (2005) Susruta Samhita, Haridas Ayurveda series 9, vol 15. Chaukhambha Visvabharati, Varanasi, Vol II, Shareera sthana, Sloka, Chapter 2, pp 128–129

Sharma PV (2007) Cakradatta, Kashi Ayurveda series 17, vol 14. Chaukhambha Orientalia, Delhi, Sloka, Chapter 33, pp 291

Sharma PV (2007) Cakradatta, Kashi Ayurveda series 17, vol 206. Chaukhambha Orientalia, Delhi, Sloka, Chapter 1, p 31

Sharma PV (2008) Dhanvantari Nighantu, Jaikrishnadas Ayurveda series 40, vol 2–3. Chaukhambha Orientalia, Varanasi, Candanadi Tritiya varga, Sloka, p 91

Sharma PV (2013) Kaiyadeva Nighantu, Jaikrishnadas Ayurveda series 30. Chaukhambha Orientalia, Varanasi, pp 232

Sharma PV (2014) Caraka Samhita, Jaikrishnadas Ayurveda series 36, vol II. Chaukhambha Orientalia, Varanasi, Chapter 3, 4,14, 19,20, Sloka pp 33,73, 79–80 86, 92, 185, 242–243, 261, 330, 338–339.

Shodhala, Gadanigraha, Rajayakshama/44 (2003) With hindi commentary 'Vidyotani' by Shri Indradev Tripathi, Shri Gangasahay Pandey (eds). Chaukhambha Sanskrit Prakashan, Varanasi, 4th edn, pp 319

Somadeva (1924) The history of sandalwood in ancient India, Kathasaritsagara (the Ocean of Story), book XII, chapter 95, ISBN-13:9789350501351. https://www.wisdomlib.org/hinduism/book/kathasaritsagara-the-ocean-of-story/d/doc363245.html

Srikanthamurthy KR (ed) (2006) Sarngadhara. Samhita. Chaukhambha Orientalia, Varanasi

Sultana A, Rahman K (2018) Review article: Santalum album Linn wood and its oil: an aromatic Unani traditional medicine with versatile pharmacological activities, Tang(Humanitas Medicine). 8(3)

Venugopalan Nair SN, Ved DK, Ravikumar K, Tabassum IF, Sureshchandra ST, … et al. (2020) Indian medicinal plants database (IMPLAD) and threatened medicinal plants of India. In: Rajasekharan P, Wani S (eds) Conservation and utilization of threatened medicinal plants. Springer, Cham. https://doi.org/10.1007/978-3-030-39793-7_3

Villiers J (2001) Great plenty of Almug trees: the trade in Southeast Asian: Aromatic woods in the Indian Ocean and China, 500 Bc-Ad 1500. The Great Circle 23 (2):24–43. http://www.jstor.org/stable/41563084. Accessed 5 March 2021

Wikipedia contributors (2021) March 3. Sandalwood. Wikipedia Free Encycl 07:49, https://en.wikipedia.org/w/index.php?title=Sandalwood&oldid=1010070924. Accessed 12 March 2021

Genomic Resources in Indian Sandalwood: Implications in Conservation and Breeding

4

Eswaran Muthulakshmi,
Chandramouli K. Madhuvanthi,
and Modhumita Ghosh Dasgupta

Abstract

Santalum album is a slow growing hemiparasitic tree which is extensively exploited due to its use in fragrance and essential oil industries. The threat to this species has reached critical level and the species is approaching commercial extinction, widening the gap between supply and demand. The genomic resources in sandalwood are limited to two whole-genome sequences and several transcriptomes. The molecular regulation of the sesquiterpene biosynthetic pathway which produces the essential oil in the heartwood has been extensively worked and most of the genes in the pathway have been functionally characterized. Hence, research in sandalwood is now poised to enable genomewide surveys to associate genetic variability with phenotypic traits, underpin adaptive potential of populations, in vitro production of essential oil through metabolite engineering, and development of super barcodes for timber forensics. The insights gained from the genomic resources in this tropical species can facilitate formulation of effective conservation and breeding strategies.

4.1 Introduction

Genomic research in forest trees has now entered a productive phase owing to the advent of low cost and high throughput next-generation sequencing platforms and 35 tree species have been assembled as reference genomes (Plomion et al. 2016). The genomic information accrued from these studies is expected to increase our understanding on development, evolution, and adaptation of tree genomes. The main objectives of tree genomics include deciphering the structure of genomes; characterizing gene functions and generation of transgenics; use of genomic platforms for accelerated breeding through marker-assisted selection (MAS) and monitoring and management of genetic resources (Grattapaglia et al. 2018; Neale et al. 2013; Neale and Kremer 2011) However, genomic studies in tree species still remains a challenge due to long generation time, large size of mature trees, late onset of economically important traits, non-availability of germplasm resources with defined genetic background, and long-term committed funding (Neale and Kremer 2011).

E. Muthulakshmi · C. K. Madhuvanthi ·
M. Ghosh Dasgupta (✉)
Institute of Forest Genetics and Tree Breeding,
R.S. Puram, Coimbatore 641002, India
e-mail: ghoshm@icfre.org

M. Gowda et al. (eds.), *The Sandalwood Genome*, Compendium of Plant Genomes,
https://doi.org/10.1007/978-3-030-93394-4_4

Santalum album L. commonly known as East Indian sandalwood or white sandalwood is a slow growing hemiroot parasitic tree belonging to the family Santalaceae. This species is indigenous to the tropical belt of the Indian peninsula, eastern Indonesia, and northern Australia (Harbaugh and Baldwin 2007). In India it is distributed over 9,600 km^2 (Radomiljac and McComb 1998; Srinivasan et al. 1992) from Kerala in the South to Uttar Pradesh in the North, in regions with varying ecoclimatic conditions and edaphic factors.

The essential oil derived from the heartwood (HW) of *Santalum* spp. is one of the major commodities in international trade due to its use in fragrance industries. The essential oil content in the HW of *S. album* is the highest (6–7% oil content) when compared to *S. spicatum* (2%) and *S. lanceolatum* (1%), explaining its strong market demand. Additionally, oil is also extracted from other species like *S. yasi*, S. *acuminatum,* and *S. austrocaledonicum* (Rao et al. 2016; Butaud et al. 2008; Howes et al. 2004). The global annual demand of sandalwood was estimated at about 5,000–6,000 MT wood and around 100 to 120 MT oil (Dhanya et al. 2010). It is predicted that the annual growth rate of global sandalwood oil market will be 6.99% and will reach USD 156 million by 2024 (https://www.marketresearchfuture.com/reports/sandalwood-oil-market-4626).

The genomic studies in sandalwood have essentially focused on understanding the molecular mechanism governing oil content, owing to its economic relevance. The whole-genome sequences in conjunction with the transcriptomic resources will facilitate trait-based breeding and conservation of genetic resources of Indian sandalwood. The molecular understanding of the sesquiterpene pathway will have industrial implications through biomarker discovery and metabolic engineering of microbes for santalol production. This chapter provides an insight into the genomic resources presently available in *Santalum* sp., with specific reference to *S. album* and its application in genetic improvement and conservation of the species.

4.2 Genomic Resources in *Santalum* Species

4.2.1 Whole-Genome Sequencing in *Santalum Album*

Indian sandalwood is one of the few tropical tree species which has been sequenced by two research groups in India. In 2018, Mahesh and coworkers (2018) used the multiomics approach to sequence the 221 Mb genome and predicted 38,119 genes. Subsequently, the second genome was reported by Dasgupta et al. (2019). Variant calling across the two genomes revealed presence of 1,191,838 (0.42%) SNPs and 98,312 (0.034%) InDels. A detailed insight into the sandal genomes will be discussed elsewhere and the present chapter focuses on the other genomic resources reported in different sandal species.

4.2.2 Organellar Genomes of *S. Album*

The significance of organellar genomes in land plants is well documented and as of June 2018, 2,484 plastid and 167 mitochondrial genomes have been deposited in NCBI (Benson et al. 2018). Variations in organellar genomes are reported to impact fitness and metabolism of plants (Joseph et al. 2013). Loss of protein-coding genes in both chloroplast and mitochondria is widely established (Gualberto et al. 2014; Stiller 2007) due to endosymbiosis and continuous inter-genomic DNA transfer between nuclear and organellar genomes (Rockenbach et al. 2016). The chloroplast genomes in land plants show high level of synteny and are structurally and functionally conserved when compared to mitochondria, which is characterized by DNA rearrangements and accumulation of heterologous DNA sequences. A typical Cp genome is arranged in a quadripartite structure, consisting of a large single copy (LSC, 80–90 kbp) region and a small single copy (SSC, 16–27 kbp) region separated by a pair of inverted repeats (IRs, 20–30 kbp) regions (Wicke et al. 2011).

The 1,44,101 bp chloroplast genome of *S. album* was reported by Yang et al. (2020). The plastome was characterized with an LSC of 83,796 bp and an SSC of 11,277 bp which is separated by 24,514 bp inverted repeats. Functional annotation predicted 123 genes, including 80 protein-coding, 8 ribosomal RNA, and 35 transfer RNA genes. Earlier, 53 chloroplast genes were mined from the whole genome of sandal (Mahesh et al. 2018), while in the later study, 117 coding genes consisting of 94 protein-coding genes, 8 rRNA genes, and 29 tRNA coding genes was reported by Dasgupta et al. (2019). The study on mitochondrial genome in sandalwood is limited to identification of 25 genes from whole-genome and transcriptome datasets (Mahesh et al. 2018). The availability of plastid genome can facilitate phylogenetic/phylogenomic analysis, phylogeographic studies, identification of novel barcodes, diversity assessment, plastid genome engineering, and gene editing.

4.2.3 Decoding of the Sesquiterpene Biosynthetic Pathway in Sandalwood

Sesquiterpenes are traditionally known to impart flavors and fragrances in the plant system. In higher plants they are synthesized in the cytosol and uses isopentenyl diphosphate (IPP) and dimethylallyl diphosphate (DMAPP) through the cytosolic mevalonate (MEV) or the plastidial methylerythritol phosphate (MEP) pathway (Srivastava et al. 2015). The essential oil from sandalwood comprises of more than 150 terpenoid compounds (Baldovini et al. 2011) which are a mixture of sesquiterpene olefins and alcohols. Approximately, 90% of essential oil comprises of (Z)-α-santalol (40–50%), (Z)-β-santalol (12–27%), (Z)-*epi*-β-santalol, and (Z)-α -exo-bergamotol (Daramwar et al. 2012; Jones, et al. 2006). Santalol biosynthesis occurs through cyclization of farnesyl diphosphate (FPP) by sesquiterpene synthases to yield sesquiterpene hydrocarbons followed by conversion to sesquiterpene alcohols by cytochrome P450 mediated hydroxylation (Jones et al. 2011).

The presence of higher levels of fragrance intensive α- and β-santalol, α -*exo*-bergamotol, and *epi*-β-santalol, (Z)-lanceol and α-bisabolol in *S. album* when compared to *S. austrocaledonicum* (Page et al. 2010) and *S. spicatum* (Piggott et al. 1997) indicates the demand of this species. Chemotype rich in sesquiterpenes like (Z)-nuciferol was reported in *S. insulare* (Butaud et al. 2008), while two chemotypes with high santalols, nuciferol, and lanceol were reported in *S. austrocaledonicum* (Bottin et al. 2007). However, no distinct chemotypes for santalol content are documented in *S. album* (Jones et al. 2006).

The first study on isolation and characterization of genes from the sesquiterpene pathway of Indian sandalwood was reported by Jones et al. (2008). Two terpene synthases (*TPS*), *SamonoTPS1* and *SasesquiTPS1* were isolated and expressed in *E. coli*. The recombinant products were functionally characterized and *SamonoTPS1* encoded (+)-α-terpineol and (−)-limonene, while *SasesquiTPS1* encoded for germacrene D-4-ol and helminthogermacrene. In a subsequent study, homologs of santalene synthase were cloned and characterized from *S. album (SaSSy)*, *S. austrocaledonicum* (*SauSSy*), and *S. spicatum* (*SpiSSy*) (Jones et al. 2011). Additionally, the genomic sequences of *Santalum TPS* genes from oil producing species (*S. album, S. spicatum, S. austrocaledonicum)* and oil deficient species (*S. murrayanum*) were compared and orthologs of santalene synthase, bisabolene/bisabolol synthase, and TPS-a sesquiterpene synthase were predicted from all species, suggesting that production of essential oil resulted from spatial and temporal expression of *TPS* genes and not from absence of orthologs or mutations in the genes (Jones et al. 2011). Subsequent reports have characterized several genes from the sesquiterpene pathway including farnesyl diphosphate synthase (*SaFDS*) and santalene synthase (*SaSS*) (Rani et al. 2013), farnesyl pyrophosphate, sesquiterpene and monoterpene synthases (Misra and Dey 2013), sesquisabinene synthases (*SaSQS1, SaSQS2*), bisabolene synthase (*SaBS*), santalene synthase (*SaSS*), and farnesyl diphosphate synthase

(*SaFDS*) (Srivastava et al. 2015), cytochrome P450 (*SaCYP736A167*) (Celedon et al. 2016), 1-deoxy-D-xylulose-5-phosphate reductoisomerase (*SaDXR*) (Zhang et al. 2021), acetyl-CoA C-acetyl transferase (*AACT*), and 3-hydroxy-3-methylglutaryl-CoA synthase (*HMGS*) (Niu et al. 2021).

The characterization of these genes also provided an insight into their biological functions in santalol biosynthesis. The study on farnesyl diphosphate synthase (*SaFDS*) and santalene synthase (*SaSS*) isolated from wood harvested from different age groups revealed higher expression of both genes in mature HW. A positive correlation between santalol content and gene expression was also established (Rani et al. 2013). Similarly, the expression levels and kinetic constants of *SaSQS1, SaSQS2, SaBS, SaSS,* and *SaFDS* isolated from the transition zone documented higher expression of *SaSS* when compared to *SaSQS1, SaSQS2,* and *SaBS* which correlated with the sesquiterpene composition of essential oil (Srivastava et al. 2015).

The sesquiterpenoid biosynthetic pathway was completed with the discovery of a multi-substrate cytochrome P450 (*SaCYP736A167*) which stereoselectively produced (Z)- α-santalol, (Z)- β-santalol, (Z)-epi-β-santalol, and (Z)- α-exo-bergamotol in heartwood tissues (Celedon et al. 2016). The study highlighted the distinct profiles of sesquiterpenols across the tissue gradient in mature wood. The HW showed a preferential expression of transcripts regulating the MEV pathway along with *FPPS* transcripts, while the sapwood (SW) had a prevalence of transcripts from the MEP pathway. The study also confirmed that biosynthesis of essential oil occurred in the HW and not in the SW or transition zone.

Other than the genes directly involved in sesquiterpene biosynthesis, upstream genes from the MVA pathway like acetyl-CoA C-acetyl transferase (*AACT*) and 3-hydroxy-3-methylglutaryl-CoA synthase (*HMGS*) were recently characterized (Niu et al. 2021). Isolation and ectopic expression of 1-deoxy-D-xylulose-5-phosphate reductoisomerase (*SaDXR*) in *Arabidopsis* indicated that *SaDXR* regulated the flux toward accumulation of sesquiterpenoids (Zhang et al. 2021).

Another potential use of essential oil from sandal is its use as insect repellent (Ritchie et al. 2006). In an effort to develop trap crop to protect tobacco from the insect pest, *Myzus persicae*, coexpression of santalene synthase (*SaSSy*) with farnesyl diphosphate synthase (*FPS*) and 3-hydroxy-3-methylglutaryl-CoA reductase (*HMGR*) or truncated *HMGR* (*tHMGR*) from *Arabidopsis thaliana* was attempted in *Nicotiana tabacum* and *N. benthamiana*. Transgenics expressing *HMGR/tHMGR* with *SaSSy* accumulated higher levels of santalenes and bergamotenes which acted as a trap for the phytophagous pest (Yin and Wong 2019). Similarly, expression of *HMGR*, *SaSSy*, and santalene/bergamotene oxidase (*SaCYP76F39v1*) also resulted in accumulation of santalenes and bergamotenes in transgenic tobacco.

The cheaper alternative of essential oil is derived from *S. spicatum* also known as Australia sandalwood (Moretta 2001). The oil is of low-grade due to presence of higher levels of E, E-farnesol (Howes et al. 2004). In an effort to underpin the sesquiterpene pathway in this species, transcriptome analysis of the xylem tissue from HW of a farnesol-rich tree was conducted (Moniodis et al. 2015). Functional annotation revealed that transcripts governing all steps of the MEV pathway except phosphomevalonate kinase were represented in the dataset, while only two enzymes from the MEP pathway were predicted. Seven *TPS* genes and several cytochromes P450, and allylic phosphatases were identified. Five new *TPS* genes were mined which belonged to the *TPS-a* and *TPS-b* subfamilies. Additionally, a multi-product sesquisabinene B synthase (*SspiTPS4*) was cloned and functionally characterized. *CYP76* was the only member in the identified cytochrome P450 oxidase and was predicted as an ortholog of *CYP76F39v1* from *S. album* with 99% identity. The study however could not explain the prevalence of E,E-farnesol in *S. spicatum* when compared to *S. album,* necessitating further studies to functionally characterize additional genes from the sesquiterpene pathway in this species.

The details of transcriptome studies in sandalwood species and the list of genes cloned and characterized from the sesquiterpene biosynthesis pathway are enumerated in Tables 4.1 and 4.2.

4.2.4 Molecular Mechanisms of Parasitism

Sandalwood is a root hemiparasite that has a specialized organ called haustoria that involves host attachment, invasion, vasculature connection, and material transfer between host and parasite. Transcriptome analysis of different stages of haustorial development in *S. album* revealed several orthologous pathways when compared with other parasitic plants like *Cuscuta pentagona* (Ranjan et al. 2014) and *Triphysaria versicolor* (Torres et al. 2005; Tomilov et al. 2005). Sandal haustorial development is regulated by genes like β-D-xylosidase, pectin lyase, and glycosyl hydrolase superfamily protein related to cell wall degradation and modification to facilitate host invasion (Zhang et al. 2015). Mitochondrial respiratory function genes were also hypothesized to be involved in haustorial development and host invasion, as energy is essential to invade host root system.

The role of phytohormones in parasite–host association is well established (Jiang et al. 2004) and also documented in sandal, where accumulation of phytohormones in the haustoria when compared to roots was reported (Zhang et al. 2012). Analysis also revealed that auxin signaling was essential for haustorial initiation, while cytokinins and gibberellins biosynthesis regulated haustorial development in sandal. Genes coding for polygalacturonase which allows parasite to soften host cell wall and penetrate the root, were also found to be up-regulated. An interesting observation in this study was the differential expression of genes involved in nodule formation like *ENODs*, *MtN3/saliva/SWEET,* and *MtN21/EamA—like/UMAMIT*. GRAS-domain transcription factors which are critical transcriptional regulators of nodule morphogenesis was also expressed during haustorial development, indicating that haustorial formation in hemiparasite like sandal may share similar molecular regulation with nodule formation in legumes (Zhang et al. 2015). This hypothesis will require further validation in other parasitic plants to establish a parallel between these two important biological events in land plants.

An interesting study on parasitism-associated plastome reduction in three facultative hemiparasites (*Pyrularia edulis, Osyris wightiana,* and *S. album*) and two obligate hemiparasites (*Viscum liquidambaricolum and V. ovalifolium*) was conducted in order Santalales. A reduction in genome size by 6–31% was observed and pseudogenization of *infA, ndhB–E, ndhG,* and *ndhK* was documented. Two genes, *ndhA* and *ndhH* were lost during evolution, indicating that plastid NDH pathway may not be essential. The reduction in Cp genome was associated with lifestyle change from photoautotroph to hemiparasitism. Evidence of lineage-specific downsizing of genome and structural rearrangement and gene loss was reported in Santalales (Guo et al. 2020).

4.2.5 Response to Abiotic Stresses

Indian sandalwood is a tropical and subtropical species and is adapted to wide range of temperature from 4.5–38 °C (Jain et al. 1998). However, low temperature is a limiting factor in introducing the species to temperate regions. In a study conducted by Zhang et al. (2017), temporal expression of cold-responsive transcripts was catalogued from leaf transcriptome datasets and 3,075 and 1,349 genes were up and downregulated, respectively. Effect of cold treatment on terpenoid pathway genes revealed upregulation of 3-hydroxy-3-methylglutaryl coenzyme A (HMG-CoA) reductase gene. Tissue-specific expression of (*E*)-β-ocimene/(*E,E*)-α –farnesene synthase (*BOS/AFS*) in leaves and santalene/bergamotene synthase 1 (*SS/BS*), *SaMonoTPS1*, and *SaCYP76F38* in roots was recorded subsequent to stress imposition. One of the important findings of the study was

Table 4.1 Summary of transcriptome analysis in different sandalwood species

Species	Sequencing platform	Tissue	Type of study	No. of contigs	No. of non-redundant contigs	No. of annotated sequences	No. of genes predicted	Reference
S. album	–	In silico analysis of previously sequenced genome and transcriptome	Genome-wide mining of WRKY family TFs and their role in abiotic stress			13,306	–	(Yan et al. 2019)
S. album	Illumina HiSeq 2000	Leaves	Response to cold stress	97,349,766	1,35,938	80,426	4,424	(Zhang et al. 2017)
S. album	Illumina HiSeq 2000	Wood	HW formation and metabolite signatures	59,000,000	5,93,510	1,11,422	5,966	(Celedon et al. 2016)
S. album	Illumina GAII Analyzer	Transition zone between HW and SW	Sesqiterpene biosynthetic pathway	26,730,799	90,476	84,094	10,668	(Srivastava et al. 2015)
S. album	Illumina HiSeq 2000	Seeds	Haustorial development	34,627,124	56,011	26,626	18,292	(Zhang et al. 2015)
S. spicatum	GSFLX 454 platform (Roche)	HW	Sesquiterpenoid biosynthetic pathway	4,89,364	12,537	–	–	(Moniodis et al. 2015)

Table 4.2 List of genes identified and characterized from the sesquiterpene biosynthetic pathway in sandalwood species

S. No	Gene ID	Gene name	Reference
S. spicatum			
1	*SspiTPS4*	sesquisabinene B synthase	(Moniodis et al. 2015)
2	*SspiBS*	a-Bisabolol synthase	(Jones et al. 2011)
3	*SpiSSy*	Santalene synthase	
4	*SspiSesquiTPS*	sesquiterpene synthase	
S. austrocaledonicum			
5	*SauSSy*	Santalene synthase	(Jones et al. 2011)
6	*SauBS*	b-bisabolene synthase	
7	*Sau-SesquiTPS*	sesquiterpene synthase	
S. murrayanum			
8	*SmSSy*	putative santalene synthase	(Jones et al. 2011)
9	*SmBS*	b-bisabolene synthase	
S. album			
10	*SaDXR*	1-deoxy-D-xylulose-5-phosphate reductoisomerase	(Zhang et al. 2021)
11	*AACT*	acetyl-CoA C-acetyl transferase	(Niu et al. 2021)
12	*HMGS*	3-hydroxy-3-methylglutaryl-CoA synthase	(Niu et al. 2021)
13	*SaTPS1*	Terpene synthase	Zhang et al. 2019)
14	*SaSSy*	Santalene synthase	(Yin and Wong 2019; Srivastava et al. 2015; Rani et al. 2013 and Jones et al. 2011)
15	*SS/BS*	santalene/bergamotene synthase 1	(Zhang et al. 2017)
16	*SaCYP76F38*	Bergamotene oxidase	(Zhang et al. 2017)
17	*SaMonoTPS1*	Monoterpene synthase	(Zhang et al. 2017; Misra and Dey 2013; Jones et al. 2008, 2011)
18	*SaCYP736A167*	Cytochromes P450	(Celadon et al. 2016)
19	*SaSQS1*	sesquisabinene synthases	(Srivastava et al. 2015)
20	*SaSQS2*		
21	*SaBS*	Bisabolene synthase	
22	*SaFDS*	Farnesyl Diphosphate synthase	(Srivastava et al. 2015; Rani et al. 2013; Jones et al. 2011)
23	*CYP76F37v1*	Bergamotene oxidase	(Diaz-Chavez et al. 2013)
24	*CYP76F37v2*		
25	*CYP76F38v1*		
26	*CYP76F38v2*		
27	*CYP76F39v1*	Sanatalene/ Bergamotene oxidase	
28	*CYP76F39v2*		
29	*CYP76F40*		
30	*CYP76F41*		
31	*CYP76F42*		
32	*CYP76F43*		
33	*FPP*	Farnesyl pyrophsophate	(Misra and Dey 2013)
34	*SaSesquiTPS1*	Sesquiterpene synthase	(Jones et al. 2008)

that ~51% of the differentially expressed genes had no annotated homologs in the Nr database. Functional characterization of these novel genes may provide a repertoire of potential genes conferring abiotic stress tolerance in tropical tree species.

In a subsequent study by the same group, the response of *TPS* to different biotic and abiotic stressors like low temperature, high light intensity, and application of methyl jasmonate (MeJA) and salicylic acid (SA) was elucidated. A total of 40 *TPS* unigenes were mined and three new homologs, *SaTPS1–SaTPS3* were reported (Zhang et al. 2019). The significant induction of *SaTPS1* and *SaTPS2* in leaves when exposed to MeJA and SA indicated their probable role during local and systemic defense response. Differential expression of all three transcripts to low temperature and high light intensity suggested their role in abiotic stress response. The function of *SaTPS3* in accumulation of essential oil in HW was also reported. The down-regulation of *SaSSY* in response to abiotic stress condition was recorded which implicates that abiotic stress conditions can adversely affect essential oil production in HW. In another study, 1,390-bp putative promoter sequence of the bisabolene synthetase (*SaBS*) gene was characterized in transgenic *Arabidopsis* and sequence analysis revealed presence of several SA-responsive TCA-elements. Histochemical GUS staining and GUS fluorometric assays revealed it as an SA-inducible promoter (Yan et al. 2020).

WRKY transcription factors are involved in response to biotic and abiotic stress, senescence, seed dormancy, and germination and also contribute to secondary metabolism (Bakshi and Oemuller 2014; Rushton et al. 2010). Genome- and transcriptome-wide analysis revealed presence of 64 WRKY TFs in *S. album*. Expression studies in callus tissues documented the induction of 25 and 24 WRKY TFs when challenged with SA and MeJA, respectively. Ectopic expression of *SaWRKY1* imparted salt tolerance in transgenic *Arabidopsis*. Further, the putative role of *SaWRKY3* and *SaWRKY49* in pathogen and salt stress responses was also implied in the study (Yan et al. 2020).

4.3 In Vivo Production and Metabolic Engineering for Santalol Production

The first proof of concept for metabolic engineering for santalol production was reported by Diaz-Chavez et al. (2013), wherein the cloning and characterization of ten *S. album* P450s belonging to CYP76F subfamily and an NADPH-dependent cytochrome P450 reductase (*CPR*) involved in santalol/bergamotol biosynthesis was reported. Expression of *SaCYP76F39v1*, *SaCPR2* and *SaSSy* in yeast cells produced (Z)- and (E)-a-santalol, (Z)- and (E)- a -exo-bergamotol, (Z)- and (E)-epi-b-santalol, and (Z)- and (E)- b santalol. Expression in bacterial system revealed that these orthologs encoded enzymes which catalyzed the formation of α-, β-, *epi*- β-santalene, and α-*exo*-bergamotene from (*E,E*)-farnesyl diphosphate (*E,E*-FPP) (Jones et al. 2011). In a recent study, microbial santalol-producing system was demonstrated (Zha et al. 2020). The expression of yeast squalene synthase (*ERG9*) was repressed and santalene synthase (*SaSS*) from *S. album* was expressed under the control of GAL promoter. The artificial santalol biosynthetic pathway was induced through over-expression of *GAL4* and *PGM2* to enhance santalol production and galactose uptake. Approximately, 1.3 g/L santalol was produced from engineered yeast cells (Zha et al. 2020). These studies hold significance for fragrance industries where metabolic engineering using these potential candidate genes can result in production of santalols and bergamotol in engineered cells.

4.4 Applications of Genomics Research in Sandal Forensics, Breeding, and Conservation

4.4.1 Timber Forensics

The demand for sandalwood timber worldwide is 6,000 MT per annum and the demand within India is 5,000 tons per annum. To meet the demand, adulterants are usually used. In 2006,

3,000 tons of *Osyris lanceolata*, was imported in the place of *S. album* (Ananthapadmanabha 2012) and as a consequence, the Government had imposed a ban on import of sandalwood in India to maintain the quality of oil. The common adulterants of sandalwood are *Osyris wightiana, Erythroxylum monogynum* (Anupama et al. 2012), and other sandalwood species which yield low-grade oil like *S. spicatum* (Rao et al. 2016; Moniodis et al. 2015; Valder et al. 2003).

Identification of adulterants in economically valuable timber is important to regulate international trade. DNA barcoding is a legally valid technique in timber forensics (Lowe and Cross 2011) and is more reliable than wood anatomy (Gasson 2011; Bhat et al. 2006), chemical profiling (Deguilloux et al. 2002), and isotopic fingerprinting (Boner et al. 2007). Discrimination of sandalwood from *O. wightiana* is difficult as the grain, color, and fragrance is highly similar (Shyaula 2012). Hence, barcodes were developed to discriminate at inter-specific and inter-generic levels. Barcodes using *rbcL, matK,* and *psbA-trnH* to distinguish sandalwood from its adulterants was reported by Dev et al. (2014). Unique SNPs in *rbcL* and *trnH-psbA* discriminated sandalwood from *E. monogynum* while barcode in *matK* differentiated the species from *O. wightiana*. In a subsequent study, combination of *psbA-trnH* + *trnK* discriminated five sandal species including *S. acuminatum*, *S. album*, *S. lanceolatum*, *S. murrayanum,* and *S. spicatum* (Jiao et al. 2019).

The major challenge of using core barcodes in species discrimination is the diagnostic power which depends on the context and can pose challenge in plants with extensive speciation due to hybridization (Gruenstaeudl et al. 2012). Further, there are no universally agreed primer binding site for the two plant barcodes *rbcL* and *matK* (Roy et al. 2010). Challenges due to PCR amplification rate and loss or duplication of corresponding loci also hampers accurate species authentication (Li et al. 2015). Hence, the PCR-free strategy of using Cp genomes reconstructed from whole-genome skimming or plastome sequencing can enable acquisition of all plastid barcodes, thus increasing the resolution (Wilkinson et al. 2017). Use of entire plastomes as super barcode also facilitates identification of hypervariable loci and lineage-specific InDels which are efficient in discriminating plant species (Fu et al. 2019; Niu et al. 2017). The Cp genome of sandalwood reported by Yang et al. (2020) and Dasgupta et al. (2019) can be used as super barcode to enhance the discriminatory power in timber forensics.

4.4.2 Biomarkers to Predict Yield and Quality of Essential Oil in HW

Gene expression is a sensitive indicator of cellular metabolism and can be considered as a "genetic signature" for genotypes, tissue types, and cellular conditions (Lettieri 2006). It has been demonstrated in humans that gene expression patterns in developmental stages are predominantly conserved (Brawand et al. 2011), suggesting that expression datasets can be used as predictive biomarkers to potentially identify the tissue status. Meta-analysis of large-scale transcriptomic datasets is a powerful tool that enables identification of core gene sets regulating complex traits (Sharifi et al. 2018). The first successful gene expression biomarker for *in planta* nitrogen status was reported in maize which was independent of genotype and developmental stage (Yang et al. 2011). Recently, meta-analysis coupled with systems biology was employed to identify tissue-specific transcriptional biomarker associated with secondary metabolism in *Echinacea purpurea* (Tahmasebi et al. 2019). Advanced wed-based system like Genevestigator can be used to perform meta-analysis of normalized and quality-controlled expression datasets (Hruz et al. 2008). Correlation of gene expression with phenotypes, mutations, or perturbations can help in modeling gene function and regulatory networks. Additionally, clustering of similar expression patterns can lead to identification of biomarkers for a given condition (Zimmermann et al. 2008). In sandalwood, the complete sesquiterpene biosynthesis pathway has been constructed and the genes governing each step have been functionally characterized. Hence, future efforts can be directed toward identification of transcriptional biomarkers to correlate with metabolite profiles in mature HW to predict the yield and quality of essential oil.

4.4.3 Population Genomics for Conservation of Sandal Genetic Resources

An increasing concern for tree biologists is the decline in genetic diversity affecting the health and resilience of populations under changing environmental conditions. Hence, documenting the diversity and its inherent dynamics is the cornerstone in genetic improvement, conservation, and sustainable utilization of tree genetic resources (Aravanopoulos 2016). It also facilitates designing of strategies and prioritizing populations for management, utilization, and conservation (Francisco-Ortega et al. 2000).

The sandalwood populations in the Indian sub-continent have reached critical level due to over-exploitation, illegal harvesting, monopoly in trade, poor germination, lack of established plantations, habitat loss, lack of regeneration due to fire and grazing, and incidence of spike disease (Dev et al. 2014; Rashkow 2014; Arun Kumar et al. 2012; 2019; Suma and Balasundaran 2003; Merlin and VanRavenswaay 1990). The species is considered to be approaching commercial extinction, necessitating conservation of genetic resources. Population genetic studies in sandal has been attempted using different marker systems like isozymes (Ratnaningrum et al. 2016; Ratnaningrum and Indrioko 2015; Rao et al. 2007; Suma and Balasundaran 2003; Angadi et al. 2003; Brand et al. 1994); RAPD (Patel et al. 2016; Dani et al. 2011; Azeez et al. 2009; Shashidhara et al. 2003; Suma and Balasundaran 2003); ISSR (Patel et al. 2016); RFLP (Jones et al. 2009) and SSR (Fatima et al. 2019; Patel et al. 2016). The populations in peninsular India were identified as the most diverse, implicating them as the "hot spot" of sandal genetic resource (Rao et al. 2007). However, there are no reports on using genomewide markers to understand the adaptive diversity of populations. In the current scenario wherein the depletion of sandal genetic resources is fast paced, it is essential to monitor and manage the adaptive variation using a strategy like landscape genomics. This will provide unprecedented information on locally adapted populations, fragile populations, and feasibility of assisted migration of gene pools across landscapes and aid in effective conservation of diverse gene pools.

4.4.4 Accelerating Trait-Based Breeding

Research in sandalwood has been carried out for more than 140 years in India but a long-term structured genetic improvement program is presently lacking in this species. Studies addressing the two commercially important traits (heartwood formation and essential oil yield) have not provided significant data to support the sandalwood improvement programs in India (Arun Kumar et al. 2012). The major bottleneck in this endeavor has been the complexity of the economically relevant traits (heartwood formation and essential oil yield), late onset of trait, unavailability of even aged harvestable populations and limited population size due to unsustainable harvest. However, there is an immense scope for genetic improvement of the species due to its predominant outcrossing nature, broad genetic base, short juvenile phase (flowering by the end of 3–4 years), flowering occurring twice a year, easy seed establishment, profuse coppicing ability, and adaptability of the species in diverse environmental and edaphic conditions. With the availability of the whole-genome and transcriptome sequences, it is now feasible to implement marker guided selection to accelerate trait-based breeding in this species. Presently, marker discovery has moved from a priori identification of discrete associations using QTL mapping and association analysis to capturing whole-genome effects using the contemporary strategy of Genomic Selection (GS) (Grattapaglia et al. 2018). The immediate target could be using this strategy to predict essential oil production in HW since the protocol for estimation of essential oil and chemical profiling is well established. The need for developing nondestructive strategies to estimate HW formation will remain a major objective in sandal research, which needs to be addressed for effective implementation of MAS.

Additional traits like biotic and abiotic tolerance also need to assessed in populations to identify tolerant genotypes and initiate trait-based breeding programs. One of the diseases of immediate concern is the sandal spike disease (SSD) caused by *Candidatus* Phytoplasma asteris. The disease is fast spreading causing significant economic loss to sandal growers. Several control measures including chemotherapy, management of insect vectors through use of insecticides and insect-exclusion screen, habitat management, and rouging have been attempted with limited success. The most effective strategy would be to breed for SSD resistance or insect vector resistance (Mondal et al. 2020). With affordability of high throughput SNP genotyping and success of GS in forestry species, it should be possible to implement MAS in sandal to ensure quality of essential oil and conserve the fast-depleting sandal genetic resources.

4.5 Conclusions

Tropical tree species like sandalwood are severely challenged by anthropogenic pressures due to changes in land-use patterns, leading to fragmented and isolated noncontiguous tree strands (Nageswara Rao et al. 2012; Gibson et al. 2011). The rapidly dwindling genetic resources in sandalwood will require immediate interventions to conserve the existing populations through a structured breeding program which can meet the demand of HW and essential oil. Genomics has provided unprecedented insight into the molecular regulation of essential oil biosynthesis and can be effectively used for in vitro production of sesquiterpenes in microbial system. Availability of whole-genome and transcriptome resources can fast pace genetic improvement, conservation, and sustainable utilization of sandal germplasm.

Acknowledgements The authors acknowledge the Ministry of Environment, Forest and Climate Change and Department of Biotechnology, Government of India for funding research programs in sandalwood. Research fellowship to E. Muthulakshmi from Department of Biotechnology and Madhuvanthi K.C. from Ministry of Environment, Forest and Climate Change is gratefully acknowledged.

References

Ananthapadmanabha HS (2012) Expert sandalwood marketing report. TFS Sandalwood Project Report 69–71

Angadi VG, Jain SH, Shankaranarayana KH (2003) Genetic diversity between sandal populations of different provenances in India. Sandalwood Res Newsl 17:4–5

Anupama C, Balasundaran M, Sujanapal P (2012) Phylogenetic relationships of *Santalum album* and its adulterants as inferred from nuclear DNA sequences. Int J Agric for 2(4):150–156

Aravanopoulos FA (2016) Conservation and monitoring of tree genetic resources in temperate forests. Curr for Rep 2:119–129

Arun Kumar AN, Joshi G, Mohan Ram HY (2012) Sandalwood: history, uses, present status and the future. Curr Sci 103:1408–1416

Arun Kumar AN, Dhyani A, Joshi G (2019) *Santalum album*. The IUCN Red List of Threatened Species 2019: e.T31852A2807668

Azeez SA, Nelson R, Prasadbabu A, Rao MS (2009) Genetic diversity of *Santalum album* using random amplified polymorphic DNAs. Afr J Biotechnol 8 (13):2943–2947

Bakshi M, Oelmüller R (2014) WRKY transcription factors: Jack of many trades in plants. Plant Signal Behav 9(2):e27700. https://doi.org/10.4161/psb.27700

Baldovini N, Delasalle C, Joulain D (2011) Phytochemistry of the heartwood from fragrant *Santalum* species: a review. Flavour Fragr J 26(1):7–26

Benson DA, Cavanaugh M, Clark K, Karsch-Mizrachi I, Ostell J, Pruitt KD, Sayers EW (2018) GenBank. Nucleic Acids Res 46(D1):D41-47

Bhat KV, Balasundaran M, Balagopalan M (2006) Identification of *Santalum album* and *Osyris lanceolata* through morphological and biochemical characteristics and molecular markers to check adulteration. KFRI Res Rep 307:19

Boner M, Sommer T, Erven C, Forstel H (2007) Stable isotopes as a tool to trace back the origin of wood. In: Proceedings of the international workshop, fingerprinting methods for the identification of timber origins. October 8–9, Bonn/Germany, pp 47–57.

Bottin L, Tassin J, Nasi R, Bouvet J (2007) Molecular, quantitative and abiotic variables for the delineation of evolutionary significant units: case of sandalwood (*Santalum austrocaledonicum* Vieillard) in New Caledonia. Conserv Genet 8(1):99–109

Brand JE, Effendi M, Fox JE (1994) Seed size and germination of *Santalum album* from Timor. Preparation. Mulga Research Center, School of Environmental Biology. Curtin University of Technology, Perth, WA6001

Brawand D, Soumillon M, Necsulea A, Julien P, Csárdi G, Harrigan P, Weier M, Liechti A, Aximu-Petri A, Kircher M (2011) The evolution of gene expression levels in mammalian organs. Nature 478(7369):343–348

Butaud JF, Raharivelomanana P, Bianchini JP, Gaydou EM (2008) *Santalum insulare* acetylenic fatty acid seed oils: comparison within the Santalum genus. J Amer Oil Chem Soc 85(4):353–356

Celedon JM, Chiang A, Yuen MM, Diaz-Chavez ML, Madilao LL, Finnegan PM, Barbour EL, Bohlmann J (2016) Heartwood-specific transcriptome and metabolite signatures of tropical sandalwood (*Santalum album*) reveal the final step of (Z)-santalol fragrance biosynthesis. Plant J 86(4):289–299

Dani KS, Ravikumar P, Kumar RP, Kush A (2011) Genetic variation within and among small isolated populations of *Santalum album*. Biol Planta. 55 (2):323–326

Daramwar PP, Srivastava PL, Priyadarshini B, Thulasiram HV (2012) Preparative separation of α-and β-santalenes and (Z)-α-and (Z)-β-santalols using silver nitrate-impregnated silica gel medium pressure liquid chromatography and analysis of sandalwood oil. Analyst 137(19):4564–4570

Dasgupta MG, Ulaganathan K, Dev SA, Balakrishnan S (2019) Draft genome of *Santalum album* L. provides genomic resources for accelerated trait improvement. Tree Genet Genom 15:34. https://doi.org/10.1007/s11295-019-1334-9.

Deguilloux MF, Premonge MH, Petit RJ (2002) Novel perspectives in wood certification and forensics: dry wood as a source of DNA. Proc Roy Soc Lond 269 (1495):1039–1046

Dev SA, Muralidharan EM, Sujanapal P, Balasundaran M (2014) Identification of market adulterants in East Indian sandalwood using DNA barcoding. Ann for Sci 71(4):517–522

Dhanya B, Syam V, Seema P (2010) Sandal (*Santalum album* L.) conservation in southern India: a review of policies and their impacts. J Trop Agri 48(1–2):1–10

Diaz-Chavez ML, Moniodis J, Madilao LL, Jancsik S, Keeling CI, Barbour EL, Ghisalberti EL, Plummer JA, Jones CG, Bohlmann J (2013) Biosynthesis of sandalwood oil: *Santalum album* CYP76F cytochromes P450 produce santalols and bergamotol. PLoS One 8(9):e75053. https://doi.org/10.1371/journal.pone.0075053

Fatima T, Srivastava A, Somashekar PV, Vageeshbabu HS, Rao MS, Bisht SS (2019) Assessment of morphological and genetic variability through genic microsatellite markers for essential oil in Sandalwood (*Santalum album* L.). 3Biotech 9(7):252. https://doi.org/10.1007/s13205-019-1758-9

Francisco-Ortega J, Santos-Guerra A, Kim SC, Crawford DJ (2000) Plant genetic diversity in the Canary Islands: a conservation perspective. Amer J Bot 87 (7):909–919

Fu CN, Wu CS, Ye LJ, Mo ZQ, Liu J, Chang YW, Li DZ, Chaw SM, Gao LM (2019) Prevalence of isomeric plastomes and effectiveness of plastome super-barcodes in yews (Taxus) worldwide. Sci Rep 9:2773. https://doi.org/10.1038/s41598-019-39161-x

Gasson P (2011) How precise can wood identification be? Wood anatomy's role in support of the legal timber trade, especially CITES. IAWA J 32(2):137–154

Gibson L, Lee TM, Koh LP, Brook BW, Gardner TA, Barlow J, Peres CA, Bradshaw CJ, Laurance WF, Lovejoy TE, Sodhi NS (2011) Primary forests are irreplaceable for sustaining tropical biodiversity. Nature 478(7369):378–381

Grattapaglia D, Silva-Junior OB, Resende RT, Cappa EP, Müller BSF, Tan B, Isik F, Ratcliffe B, El-Kassaby YA (2018) Quantitative genetics and genomics converge to accelerate forest tree breeding. Front Plant Sci 9:1693. https://doi.org/10.3389/fpls.2018.01693

Gruenstaeudl M, Santos-Guerra A, Jansen RK (2012) Phylogenetic analyses of *Tolpis Adans* (Asteraceae) reveal patterns of adaptive radiation, multiple colonization and interspecific hybridization. Cladistics 29 (2013):416–434

Gualberto JM, Mileshina D, Wallet C, Niazi AK, Weberlot F, Dietrich A (2014) The plant mitochondrial genome: dynamics and maintenance. BMC Genomics 100:107–120

Guo X, Liu C, Zhang G, Su W, Landis JB, Zhang X, Wang H, Ji Y (2020) The complete plastomes of five hemiparasitic plants (*Osyris wightiana*, *Pyrularia edulis*, *Santalum album*, *Viscum liquidambaricolum*, and *V. ovalifolium*): comparative and evolutionary analyses within Santalales. Front Genet 11:597. https://doi.org/10.3389/fgene.2020.00597

Harbaugh DT, Baldwin BG (2007) Phylogeny and biogeography of the sandalwoods (*Santalum*, Santalaceae): repeated dispersals throughout the Pacific. Amer J Bot 94(6):1028–1040

Howes MJ, Simmonds MS, Kite GC (2004) Evaluation of the quality of sandalwood essential oils by gas chromatography–mass spectrometry. J Chromatogr A 1028(2):307–312

Hruz T, Laule O, Szabo G, Wessendorp F, Bleuler S, Oertle L, Widmayer P, Gruissem W, Zimmermann P (2008) Genevestigator v3: a reference expression database for the meta-analysis of transcriptomes. Adv Bioinformatics 2008:420747. https://doi.org/10.1155/2008/420747

Jain SH, Angadi VG, Rajeevalochan AN, Shankaranarayana KH, Theagarajan KS, Rangaswamy CR

(1998) Identification of provenances of sandal in India for genetic conservation. In: Sandal and its products: Australian centre for international agricultural research (ACIAR) Proceedings, vol 84. Canberra pp 117–120

Jiang F, Jeschke WD, Hartung W (2004) Abscisic acid (ABA) flows from *Hordeum vulgare* to the hemiparasite *Rhinanthus minor* and the influence of infection on host and parasite abscisic acid relations. J Exp Bot 55(406):2323–2329

Jiao L, He T, Dormontt EE, Zhang Y, Lowe AJ, Yin Y (2019) Applicability of chloroplast DNA barcodes for wood identification between *Santalum album* and its adulterants. Holzforschung 73(2):209–218

Jones CG, Ghisalberti EL, Plummer JA, Barbour EL (2006) Quantitative co-occurrence of sesquiterpenes; a tool for elucidating their biosynthesis in Indian sandalwood *Santalum album*. Phytochemistry 67 (22):2463–2468

Jones CG, Keeling CI, Ghisalberti EL, Barbour EL, Plummer JA, Bohlmann J (2008) Isolation of cDNAs and functional characterisation of two multi-product terpene synthase enzymes from sandalwood *Santalum album* L. Arch Biochem Biophys 477(1):121–130

Jones CG, Plummer JA, Barbour EL, Byrne M (2009) Genetic diversity of an Australian *Santalum album* collection—implications for tree improvement potential. Silvae Genet 58(1–6):279–286

Jones CG, Moniodis J, Zulak KG, Scaffidi A, Plummer JA, Ghisalberti EL, Barbour EL, Bohlmann J (2011) Sandalwood fragrance biosynthesis involves sesquiterpene synthases of both the terpene synthase (TPS)-a and TPS-b subfamilies, including santalene synthases. J Biol Chem 286(20):17445–17454

Joseph B, Corwin JA, Züst T, Li B, Iravani M, Schaepman-Strub G, Turnbull LA, Kliebenstein DJ (2013) Hierarchical nuclear and cytoplasmic genetic architectures for plant growth and defence within *Arabidopsis*. Plant Cell 25(6):1929–1945

Lettieri T (2006) Recent applications of DNA microarray technology to toxicology and ecotoxicology. Environ Health Perspect 114(1):4–9

Li X, Yang Y, Henry RJ, Rossetto M, Wang Y, Chen S (2015) Plant DNA barcoding: from gene to genome. Biol Rev Camb Philos Soc 90(1):157–166

Lowe AJ, Cross HB (2011) The application of DNA to timber tracking and origin verification. IAWA J 32 (2):251–262

Mahesh HB, Subba P, Advani J, Shirke MD, Loganathan RM, Chandana SL, Shilpa S, Chatterjee O, Pinto SM, Prasad TS, Gowda M (2018) Multi-omics driven assembly and annotation of the sandalwood (*Santalum album*). Genome Plant Physiol 176 (4):2772–2788

Merlin M, VanRavenswaay D (1990) The history of human impact on the genus *Santalum* in Hawai'i. In: Proceedings of the symposium on sandalwood in the Pacific, Honolulu, Hawaii. USDA Forest Service Gen Tech Rep PSW-122, vol 122. Pacific Southwest Research Station, Forest Service, US Department of Agriculture, Berkeley, CA, pp 46–60

Misra BB, Dey S (2013) Developmental variations in sesquiterpenoid biosynthesis in East Indian sandalwood tree (*Santalum album L*). Trees 27(4):1071–1086

Mondal S, Sundararaj R, Yashavantha Rao HC (2020) A critical appraisal on the recurrence of sandalwood spike disease and its management practices. For Pathol 50(6):e12648 https://doi.org/10.1111/efp.12648

Moniodis J, Jones CG, Barbour EL, Plummer JA, Ghisalberti EL, Bohlmann J (2015) The transcriptome of sesquiterpenoid biosynthesis in heartwood xylem of Western Australian sandalwood (*Santalum spicatum*). Phytochemistry 113:79–86

Moretta P (2001) Extraction and variation of the essential oil from Western Australian sandalwood (*Santalum spicatum*). Ph. D. Thesis, The University of Western Australia, Perth, Australia

Nageswara-Rao M, Soneji JR, Sudarshana P (2012) Structure, diversity, threats and conservation of Tropical Forests. Trop For (1) https://doi.org/10.5772/39286

Neale DB, Langley CH, Salzberg SL, Wegrzyn JL (2013) Open access to tree genomes: the path to a better forest. Genome Biol 14(6):120. https://doi.org/10.1186/gb-2013-14-6-120

Neale DB, Kremer A (2011) Forest tree genomics: growing resources and application. Nat Rev Genet 12(2):111–122

Niu Z, Xue Q, Zhu S, Sun J, Liu W, Ding X (2017) The complete plastome sequences of four orchid species: insights into the evolution of the Orchidaceae and the utility of plastomic mutational hotspots. Front Plant Sci 8:715 https://doi.org/10.3389/fpls.2017.00715

Niu M, Yan H, Xiong Y, Zhang Y, Zhang X, Li Y, da Silva JAT, Ma G (2021) Cloning, characterization, and functional analysis of acetyl-CoA C-acetyltransferase and 3-hydroxy-3-methylglutaryl-CoA synthase genes in *Santalum album*. Sci Rep 11 (1):1082. https://doi.org/10.1038/s41598-020-80268-3

Page T, Southwell I, Russell M, Tate H, Tungon J, Sam C, Dickinson G, Robson K, Leakey RR (2010) Geographic and phenotypic variation in heartwood and essential-oil characters in natural populations of *Santalum austrocaledonicum* in vanuatu. Chem Biodivers 7(8):1990–2006

Patel DM, Fougat RS, Sakure AA, Kumar S, Kumar M, Mistry JG (2016) Detection of genetic variation in sandalwood using various DNA markers. 3 Biotechnology 6(1):55. https://doi.org/10.1007/s13205-016-0391-0

Piggott MJ, Ghisalberti EL, Trengove RD (1997) Western Australian sandalwood oil: extraction by different techniques and variations of the major components in different sections of a single tree. Flavour Fragr J 12 (1):43–46

Plomion C, Aury JM, Amselem J, Alaeitabar T, Barbe V, Belser C, Bergès H, Bodénès C, Boudet N, Boury C, Canaguier A (2016) Decoding the oak genome: public release of sequence data, assembly, annotation and publication strategies. Mol Ecol Res 16(1):254–265

Radomiljac AM, McComb JA (1998) Nitrogen-fixing and non-nitrogen-fixing woody host influences on the growth of the root hemi-parasite *Santalum album* L. In: Sandal and its products: Proceedings of an international seminar. Bangalore, Karnataka, India, pp. 54–57

Rani A, Ravikumar P, Reddy MD, Kush A (2013) Molecular regulation of santalol biosynthesis in *Santalum album* L. Gene 527(2):642–648

Ranjan A, Ichihashi Y, Farhi M, Zumstein K, Townsley B, David-Schwartz R, Sinha NR (2014) De novo assembly and characterization of the transcriptome of the parasitic weed dodder identifies genes associated with plant parasitism. Plant Physiol 166(3):1186–1199

Rao MN, Ganeshaiah KN, Shaanker RU (2007) Assessing threats and mapping sandal resources to identify genetic 'hot-spot'for in-situ conservation in peninsular India. Conserv Genet 8(4):925–935

Rao MS, Ravikumar G, Triveni PR, Rajan VS, Nautiyal S (2016) Analysis of policies in sustaining sandalwood resources in India. Climate change challenge (3C) and social-economic-ecological interface-building. Springer, pp 327–346

Rashkow ED (2014) Perfumed the axe that laid it low: The endangerment of sandalwood in southern India. Indian Econ Soc Hist Rev 51(1):41–70

Ratnaningrum YW, Indrioko S (2015) Response of flowering and seed production of sandalwood (*Santalum album* linn., Santalaceae) to climate changes. Procedia Environ Sci 28:665–675

Ratnaningrum YW, Faridah E, Indrioko S, Syahbudin A (2016) Flowering and seed production of sandalwood (*Santalum album*; Santalaceae) along environmental gradients in Gunung Sewu Geopark, Indonesia. Nusantara Biosci 8(2):180–191

Ritchie SA, Williams CR, Montgomery BL (2006) Field evaluation of new mountain sandalwood mosquito sticks® and new mountain sandalwood botanical repellent against mosquitoes in North Queensland, Australia. J Amer Mosq Control Assoc 22(1):158–160

Rockenbach K, Havird JC, Monroe JG, Triant DA, Taylor DR, Sloan DB (2016) Positive selection in rapidly evolving plastid-nuclear enzyme complexes. Genetics 204(4):1507–1522

Roy S, Tyagi A, Shukla V, Kumar A, Singh UM, Chaudhary LB, Datt B, Bag SK, Singh PK, Nair NK, Husain T (2010) Universal plant DNA barcode loci may not work in complex groups: a case study with Indian Berberis species. PLoS One. 2010 5(10): e13674. https://doi.org/10.1371/journal.pone.0013674

Rushton PJ, Somssich IE, Ringler P, Shen QJ (2010) WRKY transcription factors. Trend Plant Sci 15 (5):247–258

Sharifi S, Pakdel A, Ebrahimi M, Reecy JM, Farsani SF, Ebrahimie E (2018) Integration of machine learning and meta-analysis identifies the transcriptomic bio-signature of mastitis disease in cattle. PLoS One 13(2):e0191227. https://doi.org/10.1371/journal.pone.0191227

Shashidhara G, Hema MV, Koshy B, Farooqi AA (2003) Assessment of genetic diversity and identification of core collection in sandalwood germplasm using RAPDs. J Hortic Sci Biotechnol 78(4):528–536

Shyaula SL (2012) A review on genus Osyris: phytochemical constituents and traditional uses. J Nat Pharm 3(2):61–70

Srinivasan VV, Sivaramakrishnan VR, Rangaswamy CR, Ananthapadmanabah HS, Shankaranarayana KH (1992) Sandal (Santalum album Linn.). Indian council of forestry research and education (ICFRE). Dehra Dun, India.

Srivastava PL, Daramwar PP, Krithika R, Pandreka A, Shankar SS, Thulasiram HV (2015) Functional characterization of novel sesquiterpene synthases from Indian sandalwood Santalum album. Sci Rep 5 (1):10095. https://doi.org/10.1038/srep10095

Stiller JW (2007) Plastid endosymbiosis, genome evolution and the origin of green plants. Trends Plant Sci 12 (9):391–396

Suma TB, Balasundaran B (2003) Isozyme variation in five provenances of *Santalum album* in India. Aust J Bot 51(3):243–249

Tahmasebi A, Ebrahimie E, Pakniyat H, Ebrahimi M, Mohammadi-Dehcheshmeh M (2019) Tissue-specific transcriptional biomarkers in medicinal plants: application of large-scale meta-analysis and computational systems biology. Gene 691:114–124

Tomilov AA, Tomilova NB, Abdallah I, Yoder JI (2005) Localized hormone fluxes and early haustorium development in the hemiparasitic plant *Triphysaria versicolor*. Plant Physiol 138(3):1469–1480

Torres MJ, Tomilov AA, Tomilova N, Reagan RL, Yoder JI (2005) Pscroph, a parasitic plant EST database enriched for parasite associated transcripts. BMC Plant Biol 5(1):24. https://doi.org/10.1186/1471-2229-5-24

Valder C, Neugebauer M, Meier M, Kohlenberg B, Hammerschmidt F, Braun NA (2003) Western Australian sandalwood oil—new constituents of *Santalum spicatum* (R. Br.) A. DC. (Santalaceae). J Essent Oil Res 15(3):178–186

Wicke S, Schneeweiss GM, dePamphilis CW, Müller KF, Quandt D (2011) The evolution of the plastid chromosome in land plants: gene content, gene order, gene function. Plant Mol Biol 76(3–5):273–297

Wilkinson MJ, Szabo C, Ford CS, Yarom Y, Croxford AE, Camp A, Gooding P (2017) Replacing sanger with next generation sequencing to improve coverage and quality of reference DNA barcodes for plants. Sci Rep 7 (1):46040. https://doi.org/10.1038/srep46040

Yan H, Li M, Xiong Y, Wu J, Teixeira da Silva JA, Ma G (2019) Genome-wide characterization, expression profile analysis of WRKY family genes in *Santalum album* and functional identification of their role in abiotic stress. Intl J Mol Sci 20(22):5676. https://doi.org/10.3390/ijms20225676

Yan H, Xiong Y, Teixeira da Silva JA, Pang J, Zhang T, Yu X, Zhang X, Niu M, Ma G (2020) Molecular cloning and functional characterization of bisabolene synthetase (SaBS) promoter from *Santalum album*. Forests 11(1):85. https://doi.org/10.3390/f11010085

Yang XS, Wu J, Ziegler TE, Yang X, Zayed A, Rajani MS, Zhou D, Basra AS, Schachtman DP, Peng M, Armstrong CL (2011) Gene expression biomarkers provide sensitive indicators of in planta nitrogen status in maize. Plant Physiol 157(4):1841–1852

Yang D, Qiu Q, Xu L, Xu Y, Wang Y (2020) The complete chloroplast genome sequence of *Santalum album*. Mitochondrial DNA B Resour 5(1):406–407

Yin JL, Wong WS (2019) Production of santalenes and bergamotene in Nicotiana tabacum plants. PLos One 14(1):e0203249. https://doi.org/10.1371/journal.pone.0203249

Zha W, An T, Li T, Zhu J, Gao K, Sun Z, Xu W, Lin P, Zi J (2020) Reconstruction of the biosynthetic pathway of santalols under control of the GAL regulatory system in yeast. ACS Synth Biol 9(2):449–456

Zhang X, Niu M, da Silva JA, Zhang Y, Yuan Y, Jia Y, Xiao Y, Li Y, Fang L, Zeng S, Ma G (2019) Identification and functional characterization of three new terpene synthase genes involved in chemical defense and abiotic stresses in *Santalum album*. BMC Plant Biol 19(1):115. https://doi.org/10.1186/s12870-019-1720-3

Zhang Y, Yan H, Li Y, Xiong Y, Niu M, Zhang X, Teixeira da Silva JA, Ma G (2021) Molecular cloning and functional analysis of 1-Deoxy-D-Xylulose 5-Phosphate reductoisomerase from *Santalum album*. Genes 12:626. https://doi.org/10.3390/genes12050626

Zhang X, da Silva JA, Duan J, Deng R, Xu X, Ma G (2012) Endogenous hormone levels and anatomical characters of haustoria in Santalum album L. seedlings before and after attachment to the host. J Plant Physiol 169(9):859–866

Zhang X, Berkowitz O, Teixeira da Silva JA, Zhang M, Ma G, Whelan J, Duan J (2015) RNA-Seq analysis identifies key genes associated with haustorial development in the root hemiparasite Santalum album. Front Plant Sci 6:661. https://doi.org/10.3389/fpls.2015.00661

Zhang X, da Silva JA, Niu M, Li M, He C, Zhao J, Zeng S, Duan J, Ma G (2017) Physiological and transcriptomic analyses reveal a response mechanism to cold stress in Santalum album L. leaves. Sci Rep 7(1):42165. https://doi.org/10.1038/srep42165

Zimmermann P, Laule O, Schmitz J, Hruz T, Bleuler S, Gruissem W (2008) Genevestigator transcriptome meta-analysis and biomarker search using rice and barley gene expression databases. Mol Plant 1(5):851–857

5 Whole Genome Sequence of Sandalwood and Its Comparative Study

H. B. Mahesh, Meghana Deepak Shirke, K. G. Manasa, and Malali Gowda

Abstract

Sandalwood is one of the most precious tree species owing to its unique fragrance and variety of medicinal properties. In this era of advanced technologies, multi-omic datasets are available for studying a particular species. The whole genome sequences from three different geographical locations like GKVK-1 from Bangalore, Salb-1 from Kerala, and Haida-2 from China, will help to complement the proteomic and transcriptomic data sets that are available for sandalwood. The genome-wide comparison can help to analyze the variability in gene content, genome rearrangements, repeat element distribution, and genes involved in essential oil biosynthesis.

H. B. Mahesh (✉) · K. G. Manasa
Department of Genetics and Plant Breeding, College of Agriculture, Mandya, University of Agricultural Sciences, Bangalore 560065, India

M. D. Shirke
Department of Biotechnology, KTHM College, Nashik 422002, India

M. Gowda
The University of Trans-Disciplinary Health Sciences and Technology (TDU), Bengaluru 560064, India

M. Gowda
Bengaluru Genomics Center, Yelahanka, Bengaluru 560064, India

5.1 Introduction

First completed plant genome sequence of *Arabidopsis thaliana*, published in December 2000 marked the beginning of the plant genomics era. Since then multiple plant genomes have been sequenced. The short-read sequencing variants of next-generation sequencing technologies like Illumina Hiseq/NextSeq/Miseq, Roche 454 became popular platforms for whole genome sequencing of plant species. These are cost-effective, accurate, and widely compatible with various analysis tools and pipelines. The most recent development in sequencing technologies by Oxford Nanopore and PacBio have enabled long-read single molecule sequencing up to record read length as high as 23 Mb. Combination of short-read and long-read sequencing data will enable the completion of *de novo* genome assemblies, genome mapping, identification of transcript isoforms, and structural variants. In this chapter, we discuss whole genome assemblies of sandalwood genomes and their comparative analysis.

5.2 Strategies for Whole Genome Sequencing of Sandalwood Genomes

Currently there are three *Santalum album* draft genome sequences available in NCBI database; two of them are Indian sandalwood (*Santalum

M. Gowda et al. (eds.), *The Sandalwood Genome*, Compendium of Plant Genomes,
https://doi.org/10.1007/978-3-030-93394-4_5

album) species from Bangalore (GKVK-1) and Kerala (Salb-1) locations while the third *Santalum album* species (Haida-2) is from China. The two Indian cultivars have been sequenced using paired-end (PE) Illumina sequencing. GKVK-1 ecotype was sequenced using combination of paired-end (PE) and mate-pair (MP) libraries and sequenced on an Illumina HiSeq1000 platform which generated around 117 and 140 million reads, respectively. These data accounted for 95 × of sequencing depth. Kerala ecotype was sequenced using PE sequencing using Illumina NextSeq 500 and Illumina MiSeq sequencing system. Thirty-three million high-quality paired-end reads were generated which represented 30 × sequencing depth. The Haida-2 ecotype from China was sequenced using third-generation sequencing platform Oxford Nanopore with 75 × sequence depth. *Santalum album* (Indian sandalwood) has an estimated genome size of ~281 Mbp (0.29 pg). The genome size estimation of GKVK-1 was reported as 0.21 pg with consensus genome size of 221 Mb (Mahesh et al. 2018).

5.3 Sequence Library Preparation and Whole Genome Sequencing

For GKVK-1 ecotype, the DNA PE (250- to 500-bp insert size—length of the DNA fragment to be sequenced) and MP (10- to 20-kb insert size) Illumina libraries were prepared according to the instructions provided by Illumina. Libraries were sequenced using an Illumina HiSeq1000; the read length of PE was 2 × 101 bp and that of MP was 2 × 51 bp. Similarly, a strand-specific RNA sequencing library was prepared according to the instructions from the Illumina Tru-Seq RNA sample preparation kit version 2 (RS-122–2001) and sequenced (read length of 2 × 101 bp) using the Illumina HiSeq1000 platform at the Centre for Cellular and Molecular Platforms, Bengaluru, India (Mahesh et al. 2018).

For Salb-1 ecotype, genomic DNA was isolated from the young leaves using ArborEasy® DNA isolation kit (IFGTB, Coimbatore, India) (Dasgupta et al. 2019). DNA concentration and purity were estimated using NanoDrop ND-1000 UV–Vis Spectrophotometer (Thermo Scientific, MA, USA) and DNA was quantified using Qubit Fluorometer (Thermo Scientific, MA, USA). Two whole genome sequencing (WGS) libraries were prepared with Illumina-compatible NEXTflex DNA sequencing kit (BIOO Scientific, Austin, TX, USA), wherein 250 ng of genomic DNA was sheared using Covaris S2 sonicator (Covaris, Woburn, MA, USA) to generate approximate fragment size of 200 to 400 bp. The fragment size distribution was verified on Agilent 2200 TapeStation with D1000 DNA screen tapes and reagents (Agilent Technologies, Palo Alto, CA, USA) and the fragments were subsequently purified using HighPrep magnetic beads (Magbio Genomics Inc., USA). The purified fragments were end-repaired, adenylated, and ligated to Illumina multiplex barcode adaptors as suggested in NEXTFlex DNA sequencing kit protocol. The adapter-ligated DNA was purified with HighPrep beads and amplified for 10 cycles. The final PCR product was purified with HighPrep beads, followed by library quality control check. The Illumina-compatible sequencing library was quantified and 150 bp and 270 bp paired-end sequencing was done using Illumina NextSeq500 and Illumina MiSeq sequencing system. There is no detailed information on library preparation available for Haida-2.

5.4 Comparative Genomics

5.4.1 Whole Genome Assembly of Bangalore Ecotype GKVK-1

The low-quality bases (with quality score less than 30) were trimmed from raw paired-end (PE) and mate-pair (MP) sequencing reads using a FastX toolkit (http://hannonla.b.cshl.edu/fastx_toolkit/). The PE and MP reads were also trimmed for adapter sequence contamination. The good quality reads were assembled into contigs using SPAdes assembler (Bankevich et al. 2012). Contigs were assembled into scaffolds using

SSPACE (Boetzer et al. 2011) and remaining gaps were filled by the GapCloser module (Luo et al. 2012). The vector, chloroplast, and mitochondrial sequences were removed from the assembled contigs. The genome completeness was assessed using CEGMA and BUSCO (Parra et al. 2007; Simão et al. 2015).

5.4.2 Whole Genome Assembly of Kerala Ecotype Salb-1

The raw PE reads were pre-processed and quality filtered using FastQC (https://www.bioinformatics.babraham.ac.uk/projects/fastqc/). Adapter sequence contamination was removed using Cutadapt (Martin 2011). Chloroplast and mitochondrial reads were removed by reference mapping on chloroplast genome of *Osyris alba* (accession number KT070882) and mitochondrial genome of *Indica* rice (accession number DQ167399). The nuclear reads obtained after filtering the organelle reads were assembled using *de novo* assembly algorithm of CLC Genomics Workbench 8.5.2. (Dasgupta et al. 2019).

5.4.3 Whole Genome Assembly of China Ecotype Haida-2

High-quality reads of Oxford nanopore were assembled using MECAT v1.0. No further information on genome assembly and annotation is available for this genome.

5.4.4 Genome Statistics and Gene Annotation

There is considerable variation in genome assembly size of three sandalwood ecotypes (Table 5.1). The highest genome assembly length of 320,306,334 bp was obtained in case of Salb-1, followed by 220,961,086 for GKVK-1 and 196,199,524 Haida-2. The N50 for GKVK-1 was 460,663 bp, for Salb-1 was 7,958 bp and for Haida-2 was 4,363,285 bp. The average GC content was ~38% for all the three genomes. The genome completeness of GKVK-1 genome was done by Core Eukaryotic Genes Mapping Approach (CEGMA). 85% of the genes (212 out of 248) core eukaryote genes (CEGs) were found in

Table 5.1 Comparison of assembly statistics in Sandalwood ecotypes

Assembly features	GKVK-1	Salb-1	Haida-2
No. of contigs/scaffolds	12,821	61,850	180
Total length of the contigs/scaffolds (bp)	220,961,086	320,306,334	196,199,524
Minimum contig size (bp)	224	395	18,307
Maximum contig size (bp)	2,922,313	67,133	11,075,572
GC content (%)	38.38	37.89	37.22
No. of contigs with Gaps	1299	0	0
Percentage of Ns	10.13	0	0
N50 contig size (bp)	460,663	7958	4,363,285
N90 contig size (bp)	91,182	2527	704,697
Gene Features			
No. of genes predicted	38,119	62,988	28,543
Minimum gene length (bp)	153	121	102
Maximum length of gene (bp)	16,543	11,346	17,760
GC (%)	47.42	48.82	48.18
N50 length (bp)	1787	1074	1215
N90 length (bp)	459	354	456

gene repertoire of GKVK-1. The proteins sequences were annotated by BLASTP search against UniProt eudicots protein database, and annotation transfer to each gene was done by scripts bundles in Maker-P tool. Further InterProScan was used to assign the protein family (Pfam) identifier. The gene prediction resulted in 38,119 genes, among which 18,533 contained Pfam. These 18,533 genes were distributed in 3,159 types of Pfam domains. The Pfam domain containing proteins that were abundant in the sandalwood genome included protein kinase, PPR repeat family, gag-polypeptide of LTR copia type, protein Tyr kinase, RNA recognition motif, Myb-like DNA binding, cytochrome P450, zinc knuckle, ring finger, WD domain G-b repeat, plant mobile, and AP2. The Salb-1 genome annotated using GenSAS web server and gene prediction was performed using multiple tools available with GenSAS and SoftBerry to generate the consensus gene sequences as detailed by Dasgupta *et al.* Their study identified 37,500 genes with a minimum 600 nucleotide length. Eighty-five percent of these genes were validated using assembled transcripts from transcriptome reads available in NCBI SRA database. Functional annotation of the predicted genes with BlastX/Blast2Go revealed that one-third of the genes could not be assigned any functions.

5.4.5 EuKaryotic Orthologous Groups (KOG) Classification

Total of 38,119 protein sequences of GKVK-1 were subjected to KOG classification using Reverse Position Specific BLAST (RPS-BLAST) program on KOG (EuKaryotic Orthologous Groups) database with e-value 0.001 to obtain specific hit using WebMGA (Wu et al. 2011). 15,957 protein out of 38,119 were classified into 3,610 KOG families falling into 25 sub-classes into three main classes belonging to information storage and processing, cellular process and signalling, and metabolism. Kog classification of 28,543 protein of Haida-2 is done where 11,438 classified into 3,168 KOG families sharing 10,607 proteins in common and 831 unique in comparison with GKVK-1. Salb-1 is predicted to have 62,988 protein-coding genes out of them 28,906 got classified into 3,491 KOG families and sharing 14,528 proteins in common and 14,378 unique in comparison with GKVK-1. Whereas GKVK-1 has 5,350 and 1,429 unique protein-coding genes in comparison of Haida-2 and Salb-1, respectively (Fig. 5.1).

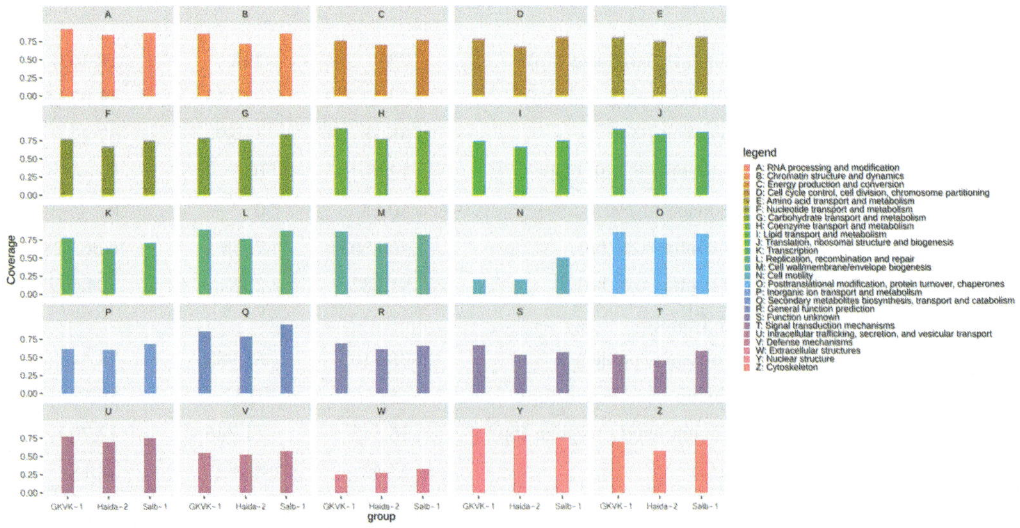

Fig. 5.1 KOG distribution across three genomes of Sandalwood

5.4.6 Structural Rearrangements of Genomes

Identification of the difference and similarities between assemblies lead us to understand the evolutionary changes and conserved sequences of the genome. NucDiff is used to find the differences and structural rearrangements among the Salb-1 and Haida-2 ecotype assemblies using GKVK-1 as reference (Table 5.2). NucDiff is capable of handling very fragmented genomes and finding the differences inside the fragments and between fragments, i.e., insertions, deletions, and substitutions, whereas structural differences between fragments covers intra- and inter-chromosomal or scaffold differences in comparison to query. Intra-chromosomal difference includes inversion, reshuffling, and relocation, whereas inter-chromosomal difference includes translocation. Haida-2 ecotype is having more fragmental differences except for substitution whereas intra- and inter-chromosomal differences are more in case of Salb-1 ecotype except for relocation in comparison to GKVK-1 as reference in both the cases.

Whole genome Average Nucleotide Identity (ANI) is estimated among three genomes using FastANI to compare and measure genomic similarity at nucleotide level. ANI estimation between Haida-2 and Salb-1 with GKVK-1 has 96% and 98% similarity, respectively. Whereas ANI estimation between Salb-1 and Haida-2 has only 94% similarities (Fig. 5.2).

5.4.7 Gene Family Expansions and Duplications

Duplication events such as whole genome duplication, segmental duplication, and local and tandem duplication are integral part of evolutionary history of flowering plants. These duplications serve as potential sources of genes with novel functions leading to contribution to species diversity. Genes of GKVK-1 (38,119), Salb-1 (62,988), and Haida-2 (28,543) were clustered using OrthoVenn2 tool. Sequence homology-based clustering with a default e-value (1e-2) and inflation value (1.5) formed 10,099 core orthologous groups (Fig. 5.3a) comprising of 43,110 genes belonging to GKVK-1 (12,234), Haida-2 (11,547), and Salb-1 (19,229) (Fig. 5.3b). Two-way comparison resulted in 3227, 4621, 1910 clusters formed between GKVK-1 and Haida-2, Salb-1and GKVK-1, Haida-2 and Salb-1, respectively. Genes which did not show homology with genes of other genomes are in-paralogs which are duplicated within genome having more than one copy. In GKVK-1 genome, 816 in-paralog clusters were identified which consists of 2591 genes. Similarly, 3898 and 129 in-paralogs were identified in Salb-1 and Haida-2 genomes, respectively (Fig. 5.3c). In addition to orthologs and in-paralogs, some genes were retained as single copy are termed as singletons which do not have sequence homology within or between species. 14,219, 11,198, and 18,190 single copy genes were identified in GKVK-1, Haida-2, and

Table 5.2 Whole genome structural differences among sandalwood genomes

Class	Haida-2 to GKVK-1	Salb-1 to GKVK-1
Insertions	350,739	200,947
Deletions	1,524,319	168,022
Substitutions	1,238,087	1,940,660
Translocations	2,184	29,270
Relocations	1,780	1,358
Inversions	175	634
Reshuffling	99	333

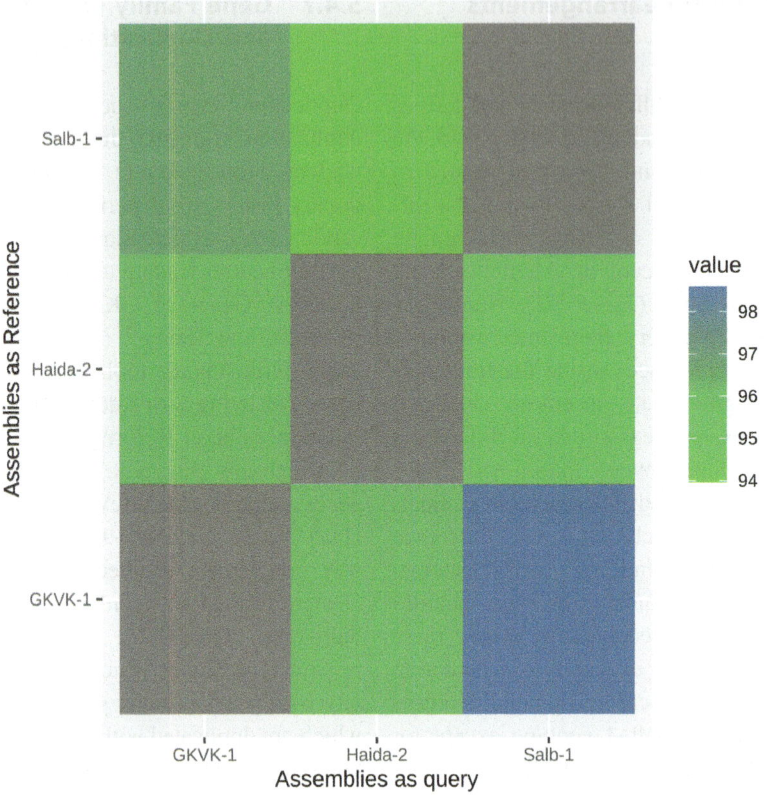

Fig. 5.2 Average nucleotide identity among three sandalwood genomes

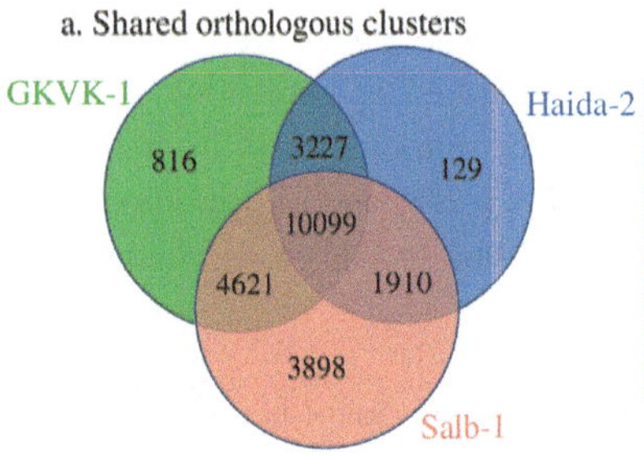

b. Counts of clusters in each genome

Genome	Proteins/Genes	Clusters	Singletons
GKVK-1	38119	18763	14219
Haida-2	28543	15365	11198
Salb-1	62988	20528	18190

Fig. 5.3 Clustering gene repertoire of Sandalwood genomes

Salb-1 genomes, respectively. Conservation of these single copy genes with respect to sequence and function are well suited for resolving phylogenetic relationships as well as to study species specificity. Similar studies were conducted in other plant species like *Arabidopsis*, *Populus*, *Vitis*, *Oryza*, and some Angiosperms (Duarte et al. 2010; Paterson et al. 2006; Han et al. 2014).

5.5 Organelle Genomes

Chloroplast and Mitochondrial genome were assembled for GKVK-1 ecotype using an ab initio method. The genome size for chloroplast 154.47 Kb and mitochondrial genome assembly size was 366.92 Kb. 53 chloroplasts and 25 mitochondrial genes were identified along with proteomic evidence for 38 (71.7%) chloroplast genes and 10 (40%) mitochondrial genes. Additionally, RNA-seq evidence was found for 8 (32%) mitochondrial genes. For Salb-1, Dasgupta et al. reported the complete chloroplast (Cp) genome assembly using *Osyris alba* Cp genome as reference. The consensus Cp genome was 147.25 Kb in length encoding 117 genes including 94 protein-coding genes, 8 rRNA genes, and 29 tRNA coding genes.

5.6 Repeats in Sandalwood Genome

Significant portions of plant genomes are occupied by repetitive DNA sequences of many distinct types. These diverse elements are difficult to analyze due to their variable nature and high copy number. The RepeatModeller was used for repeat library building and RepeatMasker was used for repeat identification. Mahesh et al. revealed 27.46% of repeats in the sandalwood genome (Table 5.3). Of 25.65% interspersed repeats, the unclassified repeats fraction was (17.28%) and retroelements (5.99%) and DNA transposons (2.38%) occupied considerably less fraction. The content of repetitive elements in Salb-1 was 12.52%, which is almost half of the GKVK-1 ecotype reported by Mahesh et al. Repeat masking revealed unclassified elements 6.93%, followed by simple repeats (4.32%). DNA transposons represented 0.14% of the total genome and included elements like hobo-Activator (0.03%), Tc1-IS630-Pogo, and Tourist/Harbinger. Retro elements occupied 1.04% of the genome and LTR elements (1.03%), LINES (0.02%), Ty1/Copia (0.69%), and Gypsy (0.33%) were also predicted in the Salb-1 draft genome (Table 5.3). The Haida-2 and GKVK-1 ecotypes showed similar statistics of repeat element distribution with 27.1% interspersed repeats, the unclassified repeats (19.41%), retroelements (5.99%), and DNA transposons (0.73%).

The considerable variation in percent repeat content between sandalwood genomes GKVK-1, Salb-1, and Haida-2 can be attributed to genome-level variation in different ecotypes, assembly tools, and differences in sequencing methods used for obtaining whole genome draft sequences. The accurate analysis of repeat elements needs third-generation sequencing platform which offer longer read length. Currently available whole genome information can be complemented with such long-range sequencing approaches to curate the genomic assemblies and accurate annotation of genomic elements.

Table 5.3 Distribution of major repeat elements in Sandalwood genomes

Repeat Class	GKVK-1	Salb-1	Haida-2*
Unclassified elements	17.28	6.93	19.41
Retroelements	5.99	1.04	6.97
SINE	0.11	0	0
LINES	1.39	0.02	1.4
LTRs	4.49	1.03	5.57
Simple repeats	0.14	4.32	0.29
DNA transposons	2.38	0.14	0.73

* Predicted using GENSAS

5.7 Comparative Analysis of Genes Involved in Santalol Biosynthesis Pathway

Genes identified in santalol biosynthesis pathway in GKVK-1 were used as reference for comparative analysis with Haida-2 and Salb-1. Santalol biosynthesis involves the Cytochrome P450 genes, Prenyltransferases, Polyprenyl synthetase, and Terpene synthetase having metal binding and N-terminal domain gene. Sequence homology search was carried out using BLASTp function of NCBI standalone blast 2.11.0 version and found minimum of 23% of identity to 100% in few genes with respect to Haida-2, whereas Salb-1 is having minimum of 30% identity in few genes and 100% identity with respect to most of the genes (Fig. 5.4).

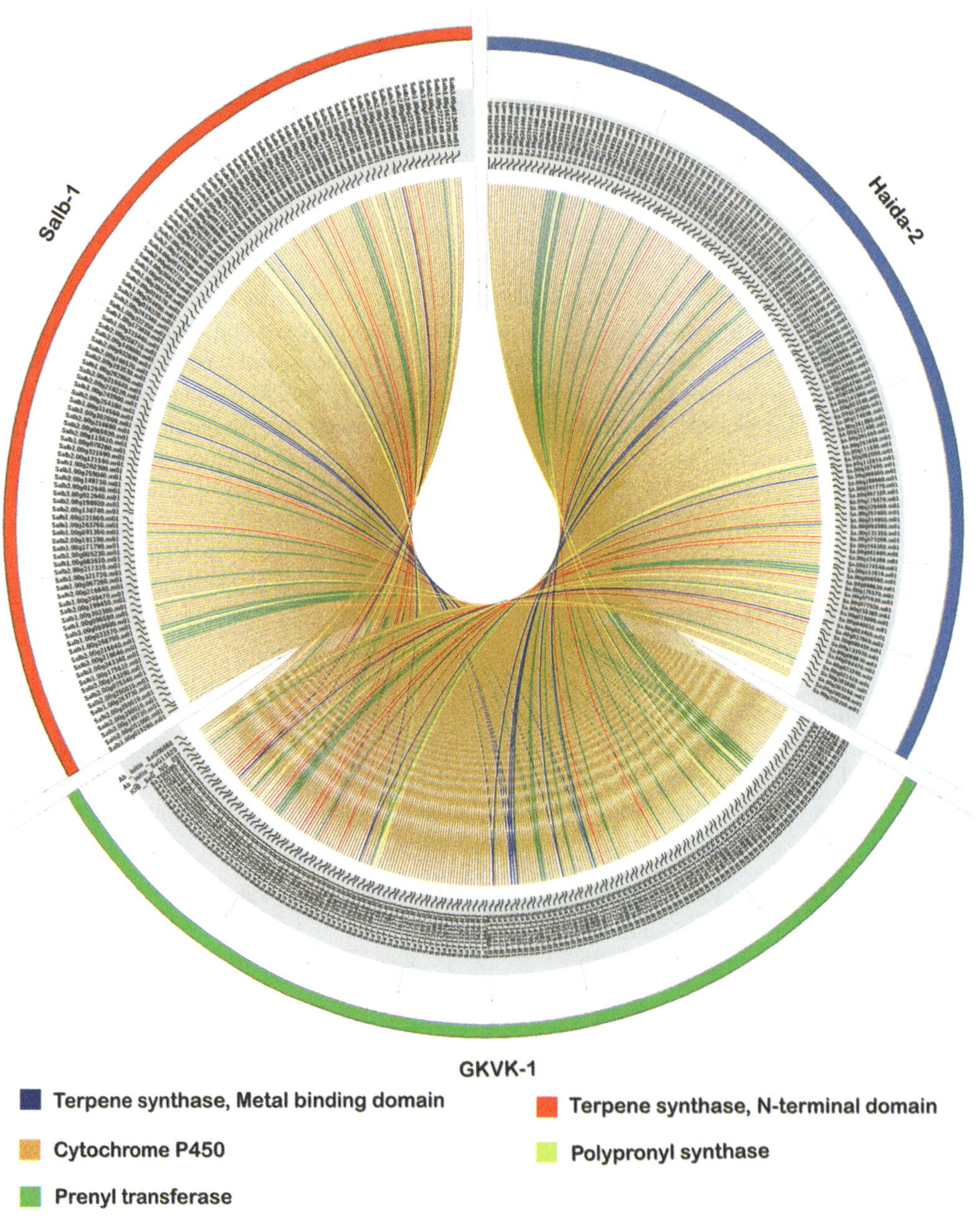

Fig. 5.4 Sequence homology of genes involved in Santalol biosynthesis in sandalwood genomes

5.8 Conclusion

The whole genome sequences of three sandalwood species are now available in the public databases. This is great resource for the ex-situ conservation efforts of sandalwood. The comparative analysis of three sandalwood genomes showed large variation in genome sizes and repeat content. Comparison of gene repertoires of three genomes helped to identify several unique and common genes. The currently available genome sequences as well as further whole genome sequencing studies of variety of germplasms will have positive impact on designing genetic markers for easy assessment of genetic diversity in sandalwood germplasms. The availability of genomic, transcriptomic, and proteomic data will help in cloning of genes involved in heartwood oil synthesis and to engineer santalol biosynthesis pathway.

References

Bankevich A, Nurk S, Antipov D, Gurevich AA, Dvorkin M et al (2012) SPAdes: a new genome assembly algorithm and its applications to single-cell sequencing. J Comput Biol 19(5):455–477

Boetzer M, Henkel CV, Jansen HJ, Butler D, Pirovano W (2011) Scaffolding pre-assembled contigs using SSPACE. Bioinformatics 27(4):578–579

Dasgupta MG, Ulaganathan K, Dev SA, Balakrishnan S (2019) Draft genome of *Santalum album* L. provides genomic resources for accelerated trait improvement. Tree Genet Genom 15 (3):1–15

Duarte JM, Wall PK, Edger PP, Landherr LL, Ma H et al (2010) Identification of shared single copy nuclear genes in Arabidopsis, Populus, Vitis and Oryza and their phylogenetic utility across various taxonomic levels. BMC Evol Biol 10(1):1–18

Han F, Peng Y, Xu L, Xiao P (2014) Identification, characterization, and utilization of single copy genes in 29 angiosperm genomes. BMC Genomics 15(1):1–9

Luo R, Liu B, Xie Y, Li Z, Huang W et al. (2012) SOAPdenovo2: an empirically improved memory-efficient short-read de novo assembler. Gigascience 1 (1):2047-2217X-2041-2018

Mahesh HB, Subba P, Advani J, Shirke MD, Loganathan RM et al (2018) Multi-omics driven assembly and annotation of the sandalwood (*Santalum album*) genome. Plant Physiol 176(4):2772–2788

Martin M (2011) Cutadapt removes adapter sequences from high-throughput sequencing reads. Embnet J 17 (1):10–12

Parra G, Bradnam K, Korf I (2007) CEGMA: a pipeline to accurately annotate core genes in eukaryotic genomes. Bioinformatics 23(9):1061–1067

Paterson AH, Chapman BA, Kissinger JC, Bowers JE, Feltus FA et al (2006) Many gene and domain families have convergent fates following independent whole-genome duplication events in Arabidopsis, Oryza, Sacchromyces and Tetraodon. Trends Genet 22 (11):597–602

Simão FA, Waterhouse RM, Ioannidis P, Kriventseva EV, Zdobnov EM (2015) BUSCO: assessing genome assembly and annotation completeness with single-copy orthologs. Bioinformatics 31(19):3210–3212

Wu S, Zhu Z, Fu L, Niu B, Li W (2011) WebMGA: a customizable web server for fast metagenomic sequence analysis. BMC Genomics 12(1):1–9

Mining of SSRs and SNPs in Sandalwood Genome

6

H. B. Mahesh, K. G. Manasa, Meghana Deepak Shirke, and N. R. Raghavendra

Abstract

Sandalwood is an economically important tree species with global demand for its fragrant heartwood oil. Considering the global demand, understanding the population diversity, genetic architecture of important traits using molecular markers may play a vital role in improving sandalwood tree genetically for long term stability and short term productivity. At the same time, deployment of molecular markers will enable understanding the genetic variation, inbreeding and gene flow. Till now, very less efforts have been undertaken to understand the population diversity of sandalwood due to lack of reliable DNA marker system. With the advent of next generation sequencing technologies, several research groups have attempted whole genome sequencing and assembly of sandalwood tree. Now, researchers can look at genetic variations at whole genome level include microsatellites/simple sequence repeats (SSRs), single nucleotide polymorphisms (SNPs), insertions and deletions (InDels) and structural variations (SVs).

H. B. Mahesh (✉) · K. G. Manasa ·
N. R. Raghavendra
Department of Genetics and Plant Breeding, College of Agriculture, Mandya, University of Agricultural Sciences, Bangalore 560065, India

M. D. Shirke
Department of Biotechnology, KTHM College, Nashik 422002, India

6.1 Introduction

Genetic improvement of forest tree species has a paramount importance but it is often neglected and overlooked research priority due to several reasons. In sandalwood, high quality heartwood, sandal oil, and spike disease resistance are the prime objectives of breeding/tree improvement. An Indian or East Indian sandalwood has great commercial value for its unique fragrant heartwood essential oil which is most preferred in perfumeries, cosmetics, medicines and incense stick industries. Ever growing global demand for unique oil, sandalwood wealth in forests are getting reduced due to overharvesting and illegal smuggling. This anthropogenic activity has put sandalwood populations under threat and conservation efforts are required for effective and sustainable use. Several such efforts have been attempted like ex situ gardening, larger scale cultivation by farming communities with a support from government. Realizing sharp decline in sandalwood population, Forest Act 1963 was amended (The Karnataka Forest (Amendment) Act, 2001) in the Karnataka Budget speech of 2001–2002, to lift the current restrictions on sandalwood plantation in private land, and to encourage public to grow sandalwood in their

M. Gowda et al. (eds.), *The Sandalwood Genome*, Compendium of Plant Genomes,
https://doi.org/10.1007/978-3-030-93394-4_6

land. Similarly, Tamil Nadu Forest (Amendment) Act 2002 has enabled legal cultivation of sandalwood. Previously, sandalwood population was threatened by pests, forest fire and smuggling that have almost denuded sandalwood cover. Due to major policy changes made by Karnataka and Tamil Nadu Governments, smuggling has drastically reduced and area under sandalwood cultivation has increased. Supply of improved true-to-type sandalwood genotypes for commercial cultivation may boost sandalwood industry. However, inadequate number of good quality seed bearing trees, well established sandalwood plantations and ever rising global demand for heartwood by oil production industries have necessitated planting of sapling in large scale. Indian Council of Forestry Research and Education has entrusted Institute of Wood Science and Technology (IWST) scientific community to promote and plant 50,000 sandalwood saplings annually. This initiative is to popularise large scale plantation of sandalwood saplings across the country and particularly in Karnataka to restore the population in natural habitat. Realizing the immense economic potential of sandalwood tree, investing on research and developmental activities of sandalwood improvement may give a paradigm shift in sandalwood conservation and sustainable use. Though conventional breeding approaches have yielded ample success to improve heartwood and essential oil production, there is a significant scope for the application of transgenics, genomics and marker assisted clonal selection in sandalwood improvement.

6.2 Molecular Markers and Their Importance in Sandalwood Improvement

Sandalwood is a less domesticated highly heterozygous and heterogeneous perennial tree species being propagated through seeds. Heartwood yield is the highly correlated trait that directly influences the oil yield. To obtain higher heartwood yield, genetically improved sandalwood breeding lines can be developed through mass clonal selection and tissue culture technique. Genetic resources, breeding, propagation and wood production are the four core elements of sandalwood breeding (Fatima et al. 2017). Fewer breeding efforts in genetic improvement of sandalwood is due to its perennial nature and lack of information on germplasm especially on high oil yielding accessions. Considering the global demand, understanding the population diversity, genetic architecture of important traits using molecular markers may play a vital role in improving sandalwood tree genetically for long term stability and short term productivity. At the same time, deployment of molecular markers will enable understanding the genetic variation, inbreeding and gene flow. Understanding the molecular mechanism and genes involved in biosynthesis of oil will fast track the breeding process to identify DNA markers which can serve as surrogate to select high oil yielding genotypes in near future. These DNA markers can be effectively deployed in marker assisted selection (MAS). However, MAS is yet to realize its potential applications in sandalwood because of several reasons like heterogeneous population, lack of trait inheritance, difficulty in development of inbred lines to understand the inheritance of quantitative traits, limited scientific community working on sandalwood. Previously, genetic diversity of *S. album* has been studied using isozyme, RAPD, ISSR and SSR-markers (Azeez et al. 2009; Dani et al. 2011; Fatima et al. 2019; Jones et al. 2009; Naseer et al. 2012; Patel et al. 2016; Rao et al. 2007; Shashidhara et al. 2003; Suma and Balasundaran 2003). However, these studies utilized comparatively fewer markers spread across sandalwood genome, due to lack of genomic resources like whole genome assembly and expressed gene sequences. With the advent of next generation sequencing (NGS) technologies, three independent research groups have sequenced the whole genome of three sandalwood ecotypes GKVK-1 (Mahesh et al. 2018), Salb1 (Dasgupta et al. 2019) and Haida2 (https://www.ncbi.nlm.nih.gov/nuccore/1338106542).
Now, researchers can look at genetic variations at whole genome level including microsatellites/ simple sequence repeats (SSRs), single

nucleotide polymorphisms (SNPs), insertions and deletions (InDels) and structural variations (SVs).

6.3 Identification of SSRs Across Multiple Sandalwood Genomes

The major hurdle in sandalwood conservation efforts is lack of information about genetic variability and sparse availability of molecular markers in sandalwood population. The availability of sandalwood genomes has generated wealth of genomic information ~~in this area~~ which will definitely assist in future studies of genetic variability. SSR and SNP markers are the choice of markers for many applications in genetic and genomic studies by virtue of their genome abundance and reproducibility of results. SSRs are the genomic regions that possess tandemly repeated nucleotide bases in a specific pattern like mono, di, tri, tetra and penta, etc. present in both coding and non-coding regions of the genome. So, we used MISA tool (Thiel et al. 2003) to identify SSRs with different repeat motifs, from mono- to hexa-nucleotide across all three genomes viz., GKVK1, Salb1 and Haida2. A total of 153, 591, 231, 301 and 139, 490 SSRs were identified in GKVK1, Salb1 and Haida2 genomes, respectively. In all three genomes, mono repeats were more predominant followed by di, tri, tetra, penta and hexa types (Fig. 6.1a). Among mono repeats, A/T type was in higher proportion (>93%) than C/G type across three genomes (Fig. 6.1b). Among A/T type, repeat number/motif varied from 10 to 59 and higher fraction of A/T types had repeat number 10–19. However, in C/G type the repeat numbers ranged from 10 to 30 (Fig. 6.1c). AT/AT type was more predominant followed by AG/CT, AC/GT and CG/CG types in di-repeats (Fig. 6.1b). In di-repeats, the frequency of repeat motif ranging from 6 to 37 and proportion of di-repeats across three genomes are shown in Fig. 6.2a. The proportion of AAT/ATT tri-repeats was in highest occurrence followed by AAG/CTT, AGG/CCT, ATC/ATG, CCG/CGG (Fig. 6.1c). The tri-repeat motif frequency across three genomes is varied from 5 to 29 and their overall distribution in three genomes is represented in Fig. 6.2b. 28 types of tetra-repeats were identified and AAAT/ATTT, ACAT/ATGT were in more proportion than other types (Figs. 6.3, 6.4 and 6.5). Penta and hexa-repeats were less in number in all the

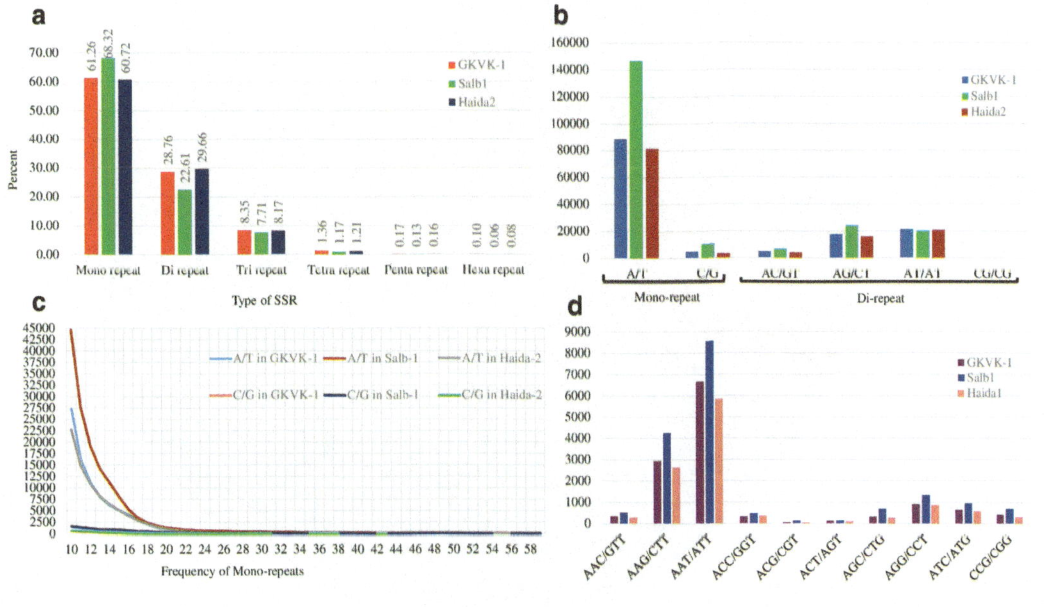

Fig. 6.1 Identification of simple sequence repeats (SSRs) in Sandalwood genome

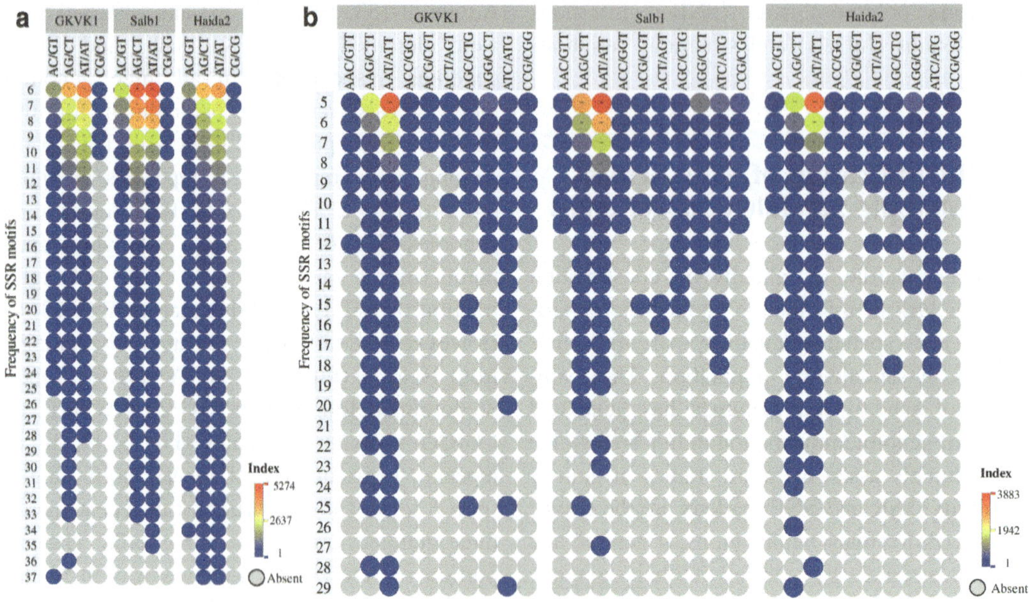

Fig. 6.2 Distribution of Di and Tri simple sequence repeats (SSRs) in Sandalwood genomes

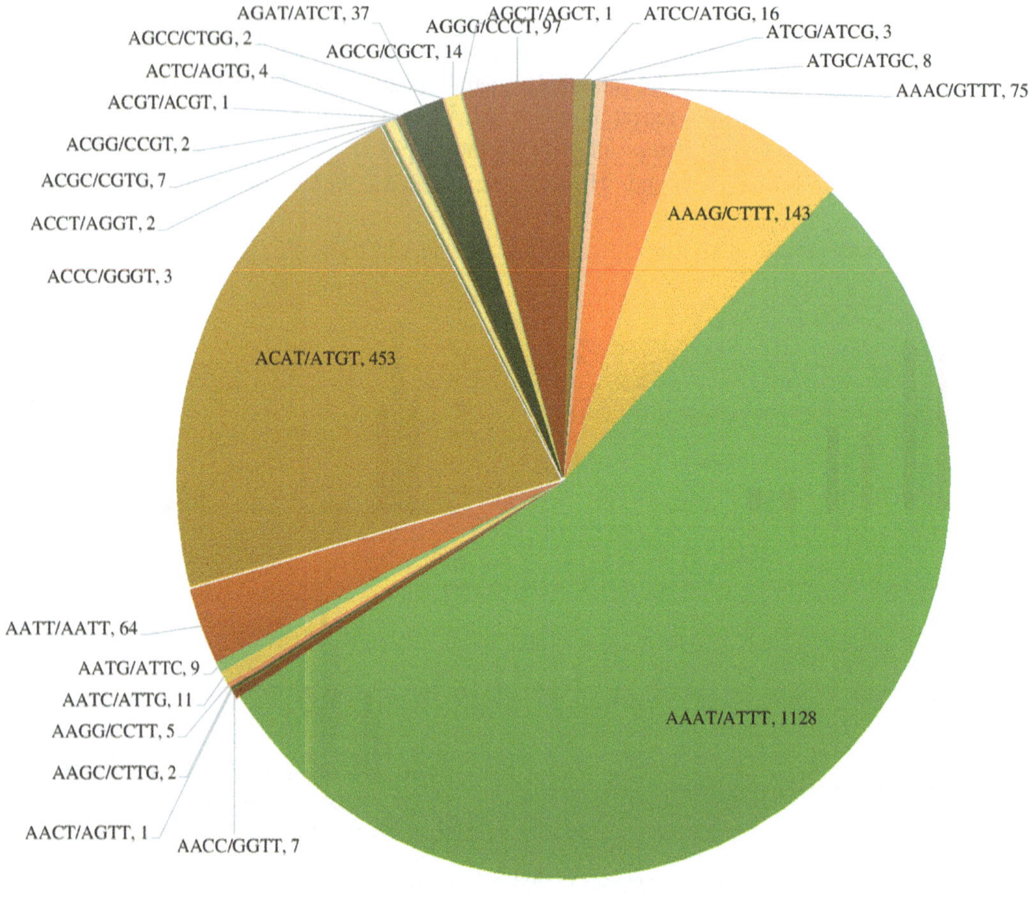

Fig. 6.3 Distribution of tetra simple sequence repeats (SSRs) in GKVK-1 Sandalwood genome

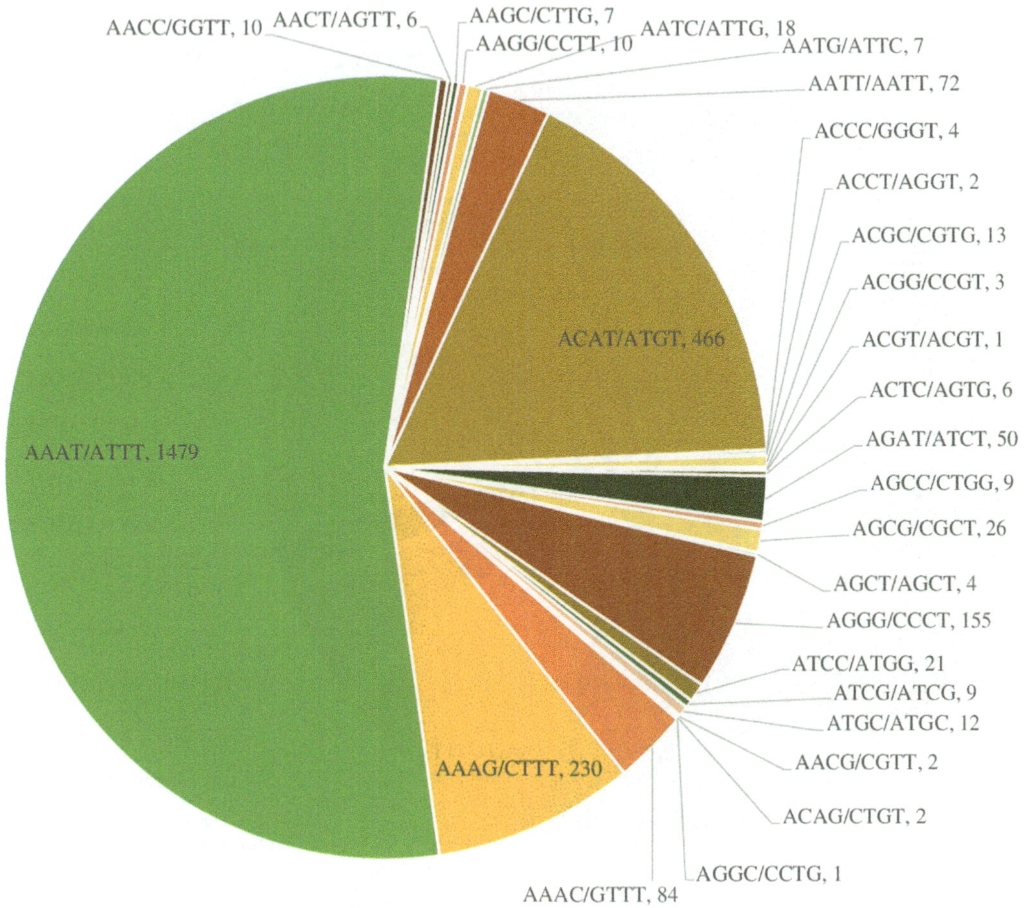

Fig. 6.4 Distribution of tetra simple sequence repeats (SSRs) in Salb-1 Sandalwood genome

genomes. Now, SSR primers can be designed to flanking conserved sequences that enable to know the variable SSR region across multiple genomes of sandalwood. Earlier fingerprinting studies (Naseer et al. 2012; Patel et al. 2016; Shashidhara et al. 2003; Suma and Balasundaran 2003) were relied on inter-specific SSRs from *S. austrocaledonicum* and *S. insulare*, RAPDs and ISSRs which have yielded ample opportunity to assess genetic variability in the sandalwood population.

6.4 Identification and Functional Annotation of SNPs

The SNPs and InDels reported between GKVK-1 and Salb-1 genomes (Dasgupta et al. 2019) from Dasgupta et al. (2019) were used for SnpEff annotation (Cingolani et al. 2012). BCFtools used to extract the basic statistics (http://samtools.github.io/bcftools/bcftools.html). Total variants reported was 11,92,739 among them 10,93,545

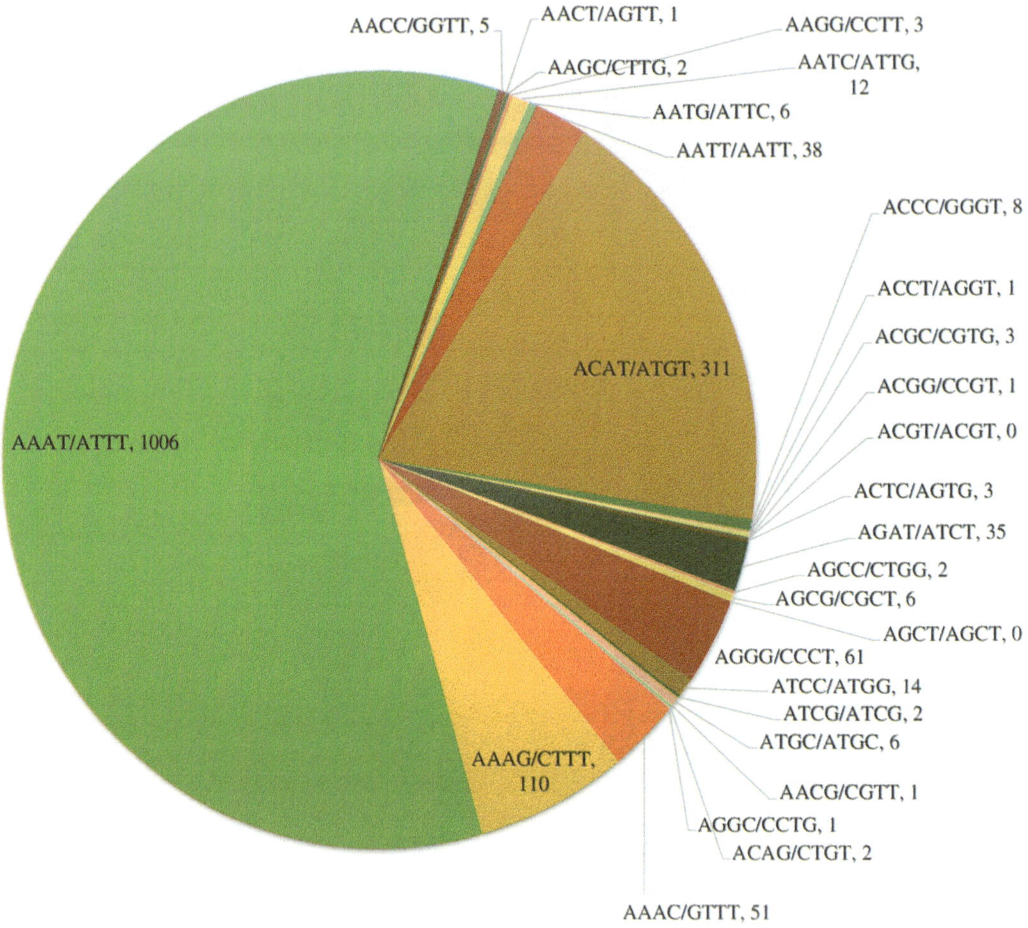

Fig. 6.5 Distribution of tetra simple sequence repeats (SSRs) in Haida-2 Sandalwood genome

single nucleotide polymorphisms (SNPs) and 99,194 were indels, among indels 52,193 were insertions of length 1 to 50 bp and 47,001 were deletions of length 1 to 39 bp (Fig. 6.6a). Among SNPs, all possible substitution of nucleotide from purine to purine, purine to pyrimidine, pyrimidine to pyrimidine and also pyrimidine to purine was observed. Purine to Purine or Pyrimidine to Pyrimidine changes are called has Transition (Ts), Purine to pyrimidine or vice versa changes are termed as Transversions (Tv). Total number of transition was found to be 958,955 and transversions were 441,68. The Ts/Tv ratio was 2.143809 (Fig. 6.6b). The Ts/Tv ratio is generally about 2, if the transversions are more and they are likely to change the amino acid and may lead to fatal mutation of translated protein. Total number of transition is found to be 9,58,955, among these we could identify purine to purine transition from A-G (2,01,830), G-A (2,04,811), remaining 4,06,705 was pyrimidine transitions, of which 2,05,995 and 2,00,710 were corresponding to T-C and C-T transition, respectively. Similar analysis revealed a total of 4,41,681 transversions with A-T, A-C, G-T, G-C, C-G, C-A, T-A and T-G as represented in Fig. 6.6c.

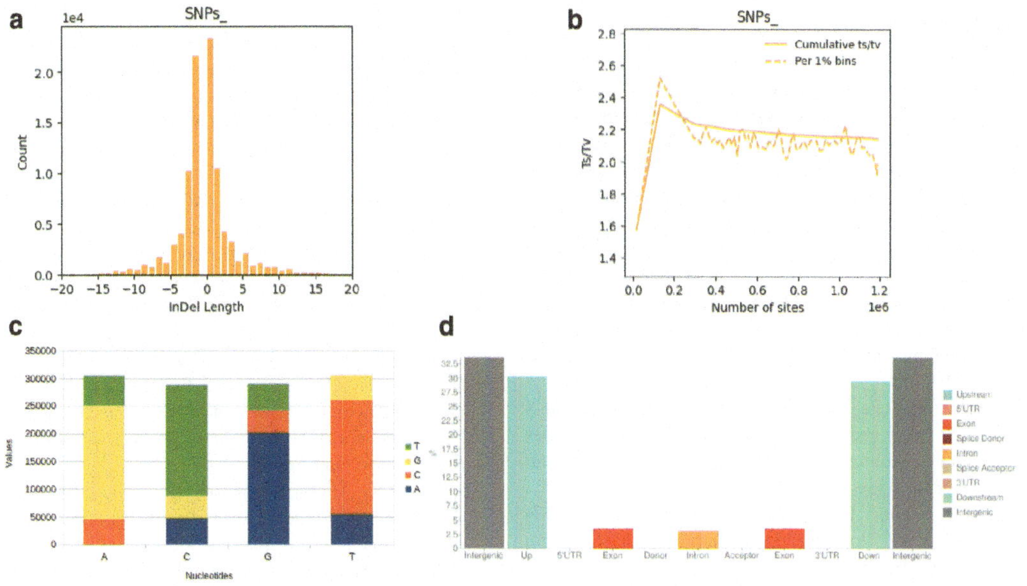

Fig. 6.6 Identification of SNPs, InDels and functional annotation of SNPs between GKVK-1 and Salb-1 genomes

SnpEff tool was used in the annotation of 11,92,739 variants which includes 10,93,545 SNPs, 52,193 insertions and 47,001 deletions. Annotation predicted the coding effects of variants which includes synonymous and non-synonymous substitution of amino acids. Among Non-synonymous substitutions, 57.61% SNPs were missense mutations and 1.63% SNPs were non-sense mutation, whereas remaining 40.75% SNPs were synonymous or silent substitutions. The maximum number of variants were intergenic (33.48%), followed by upstream variants (30.23%), downstream variants (29.44%), Intron variants (3.18%), exon variants (3.47%), untranslated regions and splice site (1%) (Fig. 6.6d). SNPs prediction and annotation was not carried out for Haida-2 genome as raw data is not accessible.

6.5 Conclusion

Sandalwood is an economically important tree species with global demand for its fragrant heartwood oil. Natural habitats of sandalwood is in declining phase due to overharvesting and illegal trading. Till now, lesser efforts have been undertaken to understand the population diversity of sandalwood due to lack of reliable DNA marker system. With the advent of next generation sequencing technologies, several research groups have attempted whole genome sequencing and assembly of sandalwood tree. Now, the sandalwood researchers can utilize this information for SSRs and SNPs based marker development for molecular analysis. These are preferred DNA markers for all molecular work like DNA fingerprinting, assessment of genetic diversity, linkage map construction, gene/QTL mapping, haplotype analysis, genome-wide association study, marker assisted selection and genomic selection. Genotyping-by-sequencing (GBS) or double digested restriction associated DNA sequencing (ddRADSeq) of natural sandalwood populations and comparing to reference whole genome sequences of sandalwood available in the public domain will be a boon to the sandalwood research community for assessment of genetic diversity and gene discovery. Identification of potential functional markers linked to specific traits and using them in sandalwood breeding/improvement programme enable transition from phenotype-based selection to genotype-based selection in near future.

References

Azeez SA, Nelson R, Prasadbabu A, Rao MS (2009) Genetic diversity of *Santalum album* using random amplified polymorphic DNAs. Afr J Biotechnol 8(13)

Cingolani P, Platts A, Wang LL, Coon M, Nguyen T et al (2012) A program for annotating and predicting the effects of single nucleotide polymorphisms, SnpEff: SNPs in the genome of *Drosophila melanogaster* strain w1118; iso-2; iso-3. Fly 6(2):80–92

Dani KS, Ravikumar P, Kumar RP, Kush A (2011) Genetic variation within and among small isolated populations of *Santalum album*. Biol Planta 55 (2):323–326

Dasgupta MG, Ulaganathan K, Dev SA, Balakrishnan S (2019) Draft genome of *Santalum album* L. provides genomic resources for accelerated trait improvement. Tree Genet Genom 15 (3):1–15

Fatima T, Hanur VS, Srivastava A, Rao MS (2017) Genetics and Biotechnology of Sandalwood (S. album L.). Intl J Sci Eng Res 8(4)

Fatima T, Srivastava A, Somashekar P, Hanur VS, Rao MS et al. (2019) Assessment of morphological and genetic variability through genic microsatellite markers for essential oil in Sandalwood (*Santalum album* L.). 3 Biotechnology 9(7):1–16

Jones C, Plummer J, Barbour E, Byrne M (2009) Genetic diversity of an Australian *Santalum album* collection–implications for tree improvement potential. Silvae Genet 58(1–6):279–286

Mahesh HB, Subba P, Advani J, Shirke MD, Loganathan RM et al (2018) Multi-omics driven assembly and annotation of the sandalwood (*Santalum album*) genome. Plant Physiol 176(4):2772–2788

Naseer M, Shashank M, Kumar P, Ansari S (2012) Genotyping of *Santalum album* L. accessions through cross-species transferability of SSR markers of *Santalum austrocaledonicum* and *Santalum insulare*. Sandalwood Res Newsl 27:1–5

Patel DM, Fougat RS, Sakure AA, Kumar S, Kumar M et al. (2016) Detection of genetic variation in sandalwood using various DNA markers. 3 Biotechnology 6 (1):55

Rao N, Ravikanth G, Ganeshaiah K, Rathore T, Shaanker R (2007) assessing threats and identifying the ecological niche of sandal resources to identify 'hot-spots' for in situ conservation in Southern India. Sandal Proc Natl Semin 23–31

Shashidhara G, Hema M, Koshy B, Farooqi A (2003) Assessment of genetic diversity and identification of core collection in sandalwood germplasm using RAPDs. J Hortic Sci Biotechnol 78(4):528–536

Suma T, Balasundaran M (2003) Isozyme variation in five provenances of *Santalum album* in India. Aust J Bot 51(3):243–249

Thiel T, Michalek W, Varshney R, Graner A (2003) Exploiting EST databases for the development and characterization of gene-derived SSR-markers in barley (*Hordeum vulgare* L.). Theor Appl Genet 106 (3):411–422

Sandalwood Fragrance Pathway and Its Engineering for Sustainable Production of High-Value Santalols

7

Ananth K. Narayanan, Sudharshan SJ, and Dinesh A. Nagegowda

Abstract

The essential oil extracted from the *Santalum* genus, known as sandalwood oil is one of the most valuable plant-derived secondary metabolites. It is obtained from the steam distillation of the heartwood of the sandalwood tree. The high value of this oil mainly stems from the limited supply, caused by difficulties in the cultivation of the plant and other socio-economic conditions. All species of the *Santalum* genus, including the highest oil-producing *Santalum album* (east Indian sandalwood), are very slow-growing that takes about 15–20 years to form heartwood from which oil is distilled. Combined with the loss of cultivatable land and poaching of wild trees make sandalwood oil very valuable. This oil has seen various applications in the aroma, perfume, cosmetic, flavor, and pharmaceutical industries. Though the essential oil of sandalwood constitutes several terpenes and terpene alcohols, the main fragrance-defining constituents of the oil are sesquiterpene alcohols (*Z*)-α-santalol, (*Z*)-β-santalol, (*Z*)-*epi*-β-santalol, and (*Z*)-α-*exo*-bergamotol, which are derived from their corresponding sesquiterpenes. In this chapter, we summarise the biosynthetic pathways involved in the formation of terpenoids and genes involved in the formation of key constituents of sandalwood essential oil. We cover different studies carried out toward the complete understanding of the pathway involved in sandalwood oil formation. We also take a look at how the deciphering of the sandalwood oil pathway has enabled bio-engineering approaches for enhanced and sustainable production using various biotechnological strategies.

A. K. Narayanan · Sudharshan SJ ·
D. A. Nagegowda (✉)
Molecular Plant Biology and Biotechnology Lab, CSIR-Central Institute of Medicinal and Aromatic Plants, Research Centre, Bengaluru 560065, India
e-mail: da.nagegowda@cimap.res.in

A. K. Narayanan · D. A. Nagegowda
Academy of Scientific and Innovative Research (AcSIR), Ghaziabad 201002, India

7.1 Introduction

Santalum album L., a member of the Santalaceae family, is one of the world's most valuable trees. It is a precious asset of the plant kingdom that is interlinked to India's history and heritage. It is commonly known as sandalwood (English), Chandan (Sanskrit/Hindi), and Srigandha (Kannada). Sandalwood, regarded as one of the oldest known perfumery materials with over 2000 years of continuous history, has maintained its eminence as a prized perfumery material from antiquity to modern times. Sandalwood has been mentioned in Indian mythology, folklore, and ancient scriptures, according to the historical

M. Gowda et al. (eds.), *The Sandalwood Genome*, Compendium of Plant Genomes,
https://doi.org/10.1007/978-3-030-93394-4_7

analysis. The sandalwood is widely considered to be indigenous to peninsular India, with a documented history dating back at least 2500 years (Sandeep and Manohara 2019). There are more than a dozen species of sandalwood in the world, the majority of which have been used to make essential oils. Among them, *S. album* (Indian sandalwood) and *Santalum spicatum* (West Australian sandalwood) are very important. Both these species grow best in a wide range of biotic and abiotic climatic conditions. In comparison to Indian sandalwood trees, Australian sandalwood trees are shorter in height and both the species' tree bark is brown to dark black. Heartwood is yellowish to dark brown and strongly scented, while the sapwood is white and odorless. In both species, leaves are thin, usually opposite and ovate (Brand et al. 1999; Dhanya et al. 2010). Flowers are purplish-brown, whereas, fruits are globose, and purple to black when ripens. The tree can grow up to 30 ft vertically and requires a minimum of 20–25 inches of annual rainfall (Brand et al. 1999; Sandeep and Manohara 2019).

The sandalwood trees are found primarily in India's drier tropical regions and the Indonesian islands of Timor and Sumba. In India, sandalwood is indigenous to the southern Indian plateaus, with the main sandalwood areas being mostly in Karnataka and adjoining districts of Maharashtra, Tamil Nadu, and Andhra Pradesh. The trees are now planted in India, China, Sri Lanka, Indonesia, Malaysia, the Philippines, and Northern Australia for commercial exploitation (Dhanya et al. 2010; Subasinghe 2013). Sandalwood oil has been used in the treatment and prevention of a wide range of ailments in traditional medicinal systems such as Ayurveda, Siddha, and Unani. Sandalwood's therapeutic and healthcare values are attributed to the high concentration of phytochemicals, particularly sesquiterpenes (Moy and Levenson 2017). Pharmacological properties ranging from antimicrobial, anti-inflammatory, and even anticancer activity of the oil have been demonstrated (Sindhu and Arora 2013; Moy and Levenson 2017).

The essential oil is extracted by steam distillation of the heartwood powder of the *S. album* tree and is highly expensive with a cost of around Rs 2,00,000–3,00,000/kg (Thomson 2020). The essential oil contains over 125 structurally related compounds, with only a few dozen having concentrations greater than 1% by weight (De Groot and Schmidt 2016, 2017). Sandalwood oil is a valuable commodity that is used in a variety of industries like perfumes, cosmetics, aromatherapy, and pharmaceuticals. The oil is listed as a natural flavoring ingredient in the Food Chemicals Codex of the United States Food and Drug Administration (FDA), and the Australian Therapeutic Goods Administration (TGA) has classified it as a Listed Medicine, making it available as an active ingredient in many non-prescription products. The International Organization for Standardization (ISO) has only issued standards for two species: *S.album* and *S. spicatum* for oil production (Moy and Levenson 2017). Of the two species, *S. album* produces an oil with much higher concentrations of alpha- and beta- santalols. The most important fragrance-defining components of sandalwood oil are (Z)-α-santalol, (Z)-β-santalol, (Z)-*epi*-β-santalol, and (Z)-α-*exo*-bergamotol and are found in abundance in the essential oil of *S. album* (Jones et al. 2011). Minor constituents of sandalwood oil include *cis*-lanceol, α-santalene, β-santalene, nuciferol, bisabolol and the sesquiterpene hydrocarbons such as α- and β-santalenes, α-bergamotenes, *epi*-β-santalene, α-, β- and γ-curcumenes, β-bisabolene, phenylpropanoids and trace amounts of alcohols, santenol, and teresantalol, aldehydes like tricycloekasantalal and isovaleradehyde, ketones like L-santenone, and acids like teresantalic acid (Baldovini et al. 2011).

While the demand for sandalwood oil is ever increasing, diseases and unsustainable exploitation of sandalwood trees have led to the decline of many natural populations. Also, unsustainable exploitation of sandalwood trees has pushed the species to the "vulnerable category" of the IUCN Red List (Dhanya et al. 2010). Sandalwood plantations are a more sustainable alternative to wild harvesting, but productivity is hampered by slow growth rates, long rotation times, a high risk of disease, and significant variation in oil yield. Alternatively, chemical synthesis of oil components, mainly santalols have been attempted, however, the quality of the synthetic fragrances was never matched to that of the naturally extracted oil and the process was uneconomical at an industrial scale (Christenson and Willis 1980; Brocke et al. 2008; Muratore et al. 2010). Other alternatives and more sustainable strategies to produce sandalwood oil include improved plantation systems, using predictive markers to develop heartwood in trees and metabolic engineering of heterologous host production systems to produce oil. These approaches rely on a complete and comprehensive understanding of the pathways and genes involved in sesquiterpenoid biosynthesis which gives sandalwood oil its characteristic properties. The biosynthetic pathway of the santalenes, santalols as well as other sesquiterpenes of sandalwood oil has been elucidated using functional genomics approach which has in turn facilitated heterologous production of sandalwood oil components (Diaz-Chavez et al. 2013; Moniodis et al. 2015; Srivastava et al. 2015; Celedon et al. 2016). This chapter summarizes the research carried out toward understanding sandalwood terpene biosynthesis including genes or enzymes involved in the formation of commercially important santalenes and santalols, and the recent advances made toward sustainable production of sandalwood oil.

7.2 Terpenoid Biosynthesis in Plants

Plants mainly synthesize their terpenes from two independent pathways, the cytosolic mevalonate pathway (MVA), and the plastidial methylerythritol phosphate (MEP) pathway (Fig. 7.1). The common intermediates arising from both these pathways are isopentyl diphosphate (IPP) and dimethylallyl diphosphate (DMAPP) (Kuzuyama and Seto 2012; Gutensohn et al. 2013). The head-to-tail arrangement of these five-carbon molecules gives rise to the basic skeleton of almost all terpenes seen in plants (Vranová et al. 2013; Zerbe and Bohlmann 2015). The MVA pathway is said to be responsible for the sesquiterpenoids and triterpenoids while the MEP pathway is mostly responsible for monoterpene production in plants (Wang et al. 2019).

7.2.1 The Mevalonate Pathway

The MVA pathway is a well-studied pathway for terpene biosynthesis (Tholl 2015). It is more or less conserved throughout evolution as it is found in archaea, some bacteria, and eukaryotes (Lombard and Moreira 2011). The pathway begins with the condensation of two acetyl-CoA molecules to form acetoacetyl-CoA, catalyzed by acetoacetyl-CoA thiolase (EC 2.3.1.9). This is followed by another condensation to 3-hydroxy-3-methyl-glutaryl-CoA (HMG-CoA) catalyzed by HMG-CoA synthase (EC 4.1.3.5) (Fig. 7.1). HMG-CoA is then reduced to yield mevalonate by the enzyme HMG-CoA reductase. HMG-CoA reductase is the rate-limiting enzyme in the MVA pathway and plays a key role in the formation of terpenoids in plants (Vranová et al. 2013; Liao et al. 2014). The conversion of mevalonate to IPP has variants in some archaea but in plants, mevalonate is phosphorylated twice, firstly by mevalonate kinase (2.7.1.36) to form phosphomevalonate and then by a phosphomevalonate kinase (EC 2.7.4.2) to form mevalonate pyrophosphate. Finally, the mevalonate pyrophosphate is decarboxylated to yield IPP by mevalonate pyrophosphate decarboxylase (EC 4.1.1.33). IPP is further isomerized to its isomer dimethylallyl diphosphate (DMAPP) and vice versa by the action of IPP isomerase (IPPI) (Nagegowda and Gupta 2020) (Fig. 7.1).

7.2.2 The Methylerythritol Phosphate Pathway

The initial step in the MEP pathway is the formation of 1-deoxy-D-xylulose 5-phosphate (DXP) by the condensation of pyruvate and D-glyceraldehyde 3-phosphate (G3P), and this reaction is catalyzed by DXP synthase (DXS, EC 2.2.1.7) (Fig. 7.1). This enzyme has a typical

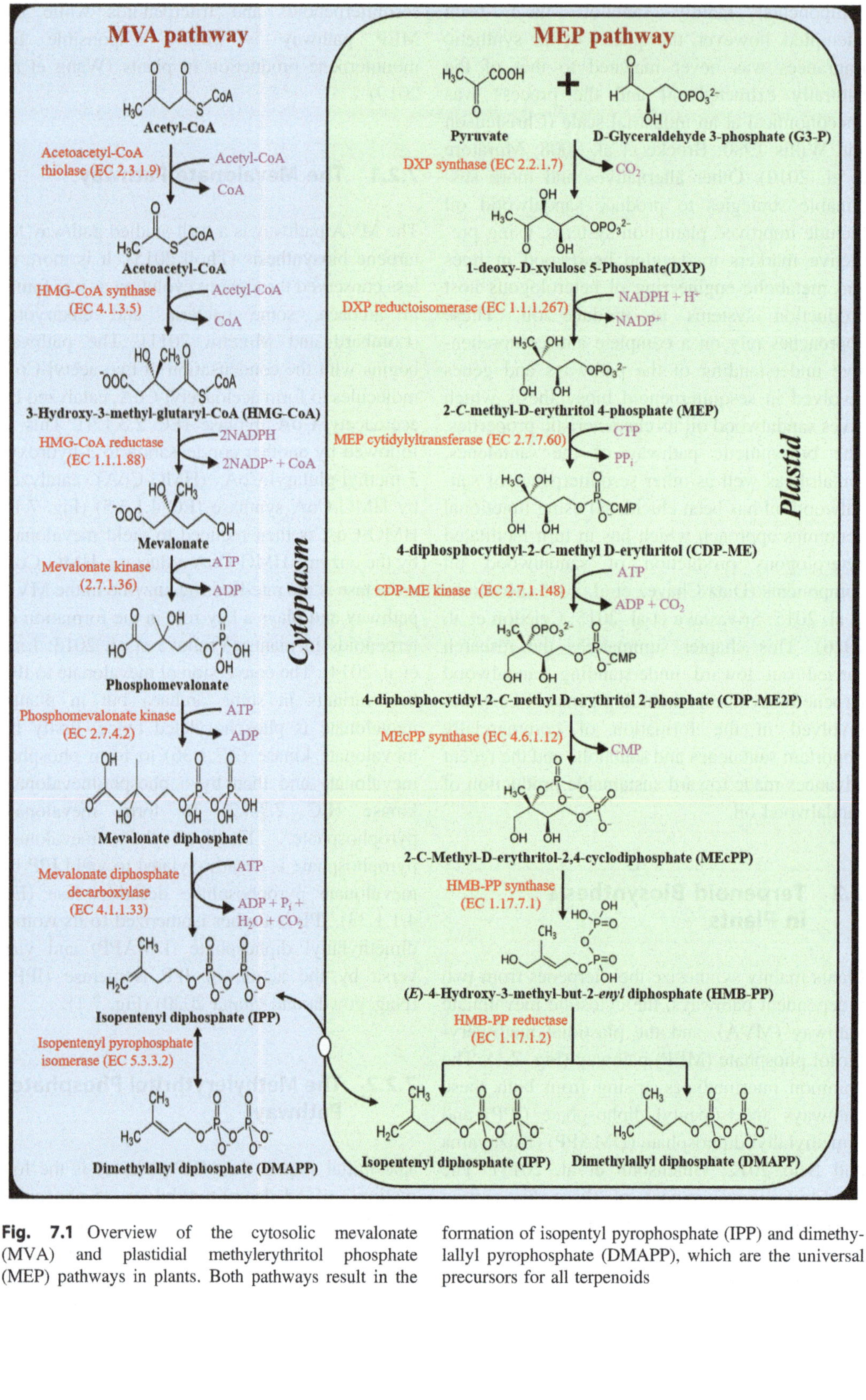

Fig. 7.1 Overview of the cytosolic mevalonate (MVA) and plastidial methylerythritol phosphate (MEP) pathways in plants. Both pathways result in the formation of isopentyl pyrophosphate (IPP) and dimethylallyl pyrophosphate (DMAPP), which are the universal precursors for all terpenoids

thiamine-binding motif and requires both thiamine and a divalent cation, such as Mg^{2+} or Mn^{2+}, for enzyme activity. After this DXP is reduced and rearranged to form 2-*C*-methylerythritol 4-phosphate (MEP) by the enzyme DXP reductoisomerase (DXR, EC 1.1.1.267) in the presence of NADPH and divalent metal ions. This is the first committed step of the MEP pathway in which a key intermediate, MEP is formed. The next step is the conversion of MEP into 4-(cytidine 5′-diphospho)-2-*C*-methyl-d-erythritol (CDP-ME) by the enzyme MEP cytidylyltransferase (MCT, EC 2.7.7.60) using CTP as a cofactor. This is followed by the phosphorylation of CDP-ME to form 2-phospho-4-(cytidine 5′-diphospho)-2-*C*-methyl-D-erythritol (CDP-ME2P) which is catalyzed by the enzyme CDP-ME kinase (CMK, EC 2.7.1.148). CDP-ME2P is then converted to 2-*C*-methyl-D-erythritol 2,4-cyclodiphosphate (MEcPP) by the enzyme MEcPP synthase (MDS, EC 4.6.1.12). The next reaction in the pathway is the conversion of MEcPP to 4-hydroxy-3-methyl but-2-enyl diphosphate (HMB-PP) by HMB-PP synthase (HDS, EC 1.17.7.1) and finally reduction of HMB-PP to IPP and dimethylallyl pyrophosphate (DMAPP) by the enzyme HMB-PP reductase (HDR, EC 1.17.1.2) (Rohdich et al. 2001; Eisenreich et al. 2004; Seemann et al. 2006; Nagegowda 2010; Zhao et al. 2013; Nagegowda and Gupta 2020) (Fig. 7.1).

7.2.3 Formation of Prenyl Diphosphate Precursors

IPP and DMAPP attach and rearrange to form the basic carbon skeleton of all terpenes found in plants. IPP and DMAPP can be interconverted thanks to the enzyme isopentyl pyrophosphate isomerase (EC 5.3.3.2). One molecule each of IPP and DMAPP are condensed to form the 10-carbon molecule geranyl diphosphate (GPP), catalyzed by the enzyme GPP synthase (GPPS) (EC 2.5.1.1) (Fig. 7.1). GPP serves as the main precursor for monoterpene formation in plants. GPPSs in plants are known to be localized in plastids and exist as homodimeric and heterodimeric structures composed of two identical sub-units or one small subunit (SSU) with one large subunit (LSU), respectively (Nagegowda 2010). The condensation of one molecule of DMAPP with two molecules of IPP generates the 15-carbon molecule farnesyl diphosphate (FPP), which is catalyzed by FPP synthase (FPPS) (EC 2.5.1.92) (Kellogg and Poulter 1997; Nagegowda and Dudareva 2008). FPP is the direct precursor for most sesquiterpenes and will be the main focus in this chapter when discussing the formation of sandalwood sesquiterpenes (Fig. 7.1). Generally, FPPS is localized in the cytosol and provides a precursor for sesquiterpenes, triterpenes, and phytosterols. However, plant genomes seem to encode various FPPS isoforms that are also localized to mitochondria, peroxisomes, and plastids (Nagegowda and Gupta 2020). For instance, a *z*FPPS localized to plastids of the glandular trichomes in wild tomatoes is involved in the biosynthesis of sesquiterpene volatiles (Sallaud et al. 2009). Diterpenes are formed from geranylgeranyl pyrophosphate (GGPP) which itself is synthesized by the catalysis of GGPP synthase (GGPPS) (EC 2.5.1.29) from DMAPP and three molecules of IPP (Fig. 7.1). Isoforms of this enzyme have been seen in the endoplasmic reticulum, plastids, and mitochondria (Jia and Chen 2016). Triterpenes are formed from the precursor squalene (SQ), which is in turn formed from the condensation of two FPP molecules by the action of the enzyme squalene synthase (SQS) (EC 2.5.1.21) (Miettinen et al. 2017) (Fig. 7.1). The formation of terpenes generally follows the pathways described above but exceptions do exist. Some examples include the non-head-to-tail condensation of two DMAPP units to form chrysanthemyl diphosphate (C10; CPP) and lavandulyl diphosphate (C10; LPP) catalyzed by CPP synthase (CPPS) and LPP synthase (LPPS), respectively (Rivera et al. 2001; Demissie et al. 2013). ChPP and LPP act as intermediates in the formation of irregular monoterpenoids pyrethtrin and lavandulol, respectively. Another example is the use of nerylneryldiphosphate (NNPP) instead of GGPP as a substrate to form certain diterpenes in tomatoes. NNPP is catalyzed by NNPP synthase (Akhtar et al. 2013).

7.3 Sandalwood Oil Biosynthesis

Sandalwood oil (hereafter, SO) is obtained by the steam distillation of the heartwood of the sandalwood tree. Its composition varies greatly with the species, region of growth, and age of the tree. SO is a complex mixture of plant secondary metabolites including terpenes and olefins. The major constituents of SO are sesquiterpene alcohols, (*Z*)-α-santalol (~50–60%), and (*Z*)-β-santalol (~20–25%) where α-santalol is usually twice the concentration of β-santalol. It is also said that these two components impart the characteristic aroma of SO and are in high demand in the perfumery industry. In addition, SO also has minor amounts of (*Z*)-*epi*-β-santalol and (*Z*)-α-*exo*-bergamotol and trace concentrations of santalenes, bergamotenes, curcumenes, bisabolenes, and their various isomers (Burdock and Carabin 2008; Baldovini et al. 2011).

The formation of terpenes/terpenoids from their prenyl diphosphate precursors, GPP, FPP, GGPP (Fig. 7.2) are often catalyzed by terpene synthases (TPS) and further modified by cytochrome P450 monooxygenases (CYP450) (Fig. 7.2) (Chen et al. 2011). One of the first studies on sandalwood TPSs was the functional characterization of SamonoTPS1 and SasesquiTPS1, however, the terpenes produced by the protein product of these genes, were only minor products found in SO (Jones et al. 2008). Later on, santalene/bergamotene synthases, which convert FPP into a blend of α-santalene, β-santalene, *epi*-β-santalene, and α-*exo*-bergamotene were identified and functionally characterized from three species of sandalwood including *S. album* (SaSSY), *S. spicatum* (SspiSSY), and *S. austroceladonicum* (SausSSY) (Jones et al. 2011) (Fig. 7.2). Extensive analysis of *S. spicatum* transcriptome showed that transcripts for the MVA and MEP pathway genes as well as five previously uncharacterized TPSs and several cytochrome P450 enzymes were well represented in the transcriptome (Moniodis et al. 2015). Characterization of one of this unknown TPSs led to the discovery of a multiproduct sesquisabinene synthase (SspiTPS4) (Moniodis et al. 2015). Other transcriptome-related studies in *S. album* led to the identification and characterization of TPSs that included sesquisabinene synthases (SaSQS1, SaSQS2), bisabolene synthase (SaBS), nerolidol/linalool synthase (SaNES/LIS), α-terpineol synthase, β-farnesene/nerolidol synthase as well as a FPPS (SaFPPS) (Srivastava et al. 2015; Zhang et al. 2019, 2021). Santalene synthases have also been characterized in non-sandalwood species such as *Clausena lansium* (SanSyn) and Camphor tree *Cinnamomum camphora* (CiCaSSY) (Di Girolamo et al. 2020). Interestingly, CiCaSSY had 97% sequence similarity with a monoterpene synthase, myrcene synthase (CiCaMS) in the same tree. This prompted further investigation into the role of amino acid residues in the formation of sesquiterpenes. They found that mutation of six amino acid residues in CaCiMS, caused the enzyme to produce sesquiterpenes α-santalene, β-santalene, and trans-α-bergamotene, the same as CaCiSSY in vitro assays with FPP as substrate (Di Girolamo et al. 2020). All these above-mentioned TPSs cyclise FPP to form α-santalene and β-santalene as well as other isomers in small quantities. Studies on the amino acid sequence of the sandalwood TPSs revealed the presence of a conserved N-terminus $R(R/P)X_8W$ motif which is crucial for the cyclization of FPP (Bohlmann et al. 1998; Chen et al. 2011). Additionally, they also contained the DDXXD motif at the C-terminus which is essential for metal-dependent ionization of the prenyl diphosphate substrate. Analysis of the sandalwood transcriptome and peptide sequence has predicted that FPPS and sesqui-TPSs could be present in the cytosol as they lack any transit peptide (Chen et al. 2011; Jones et al. 2011; Zhang et al. 2019).

Santalols are formed by the hydroxylation of santalenes at the twelfth carbon by the action of CYPs. Previous studies have characterized nine CYPs from *S. album* that are capable of carrying out this reaction and producing α-santalol, β-santalol, *exo*-α-bergamotol, and *epi*-β-santalol.

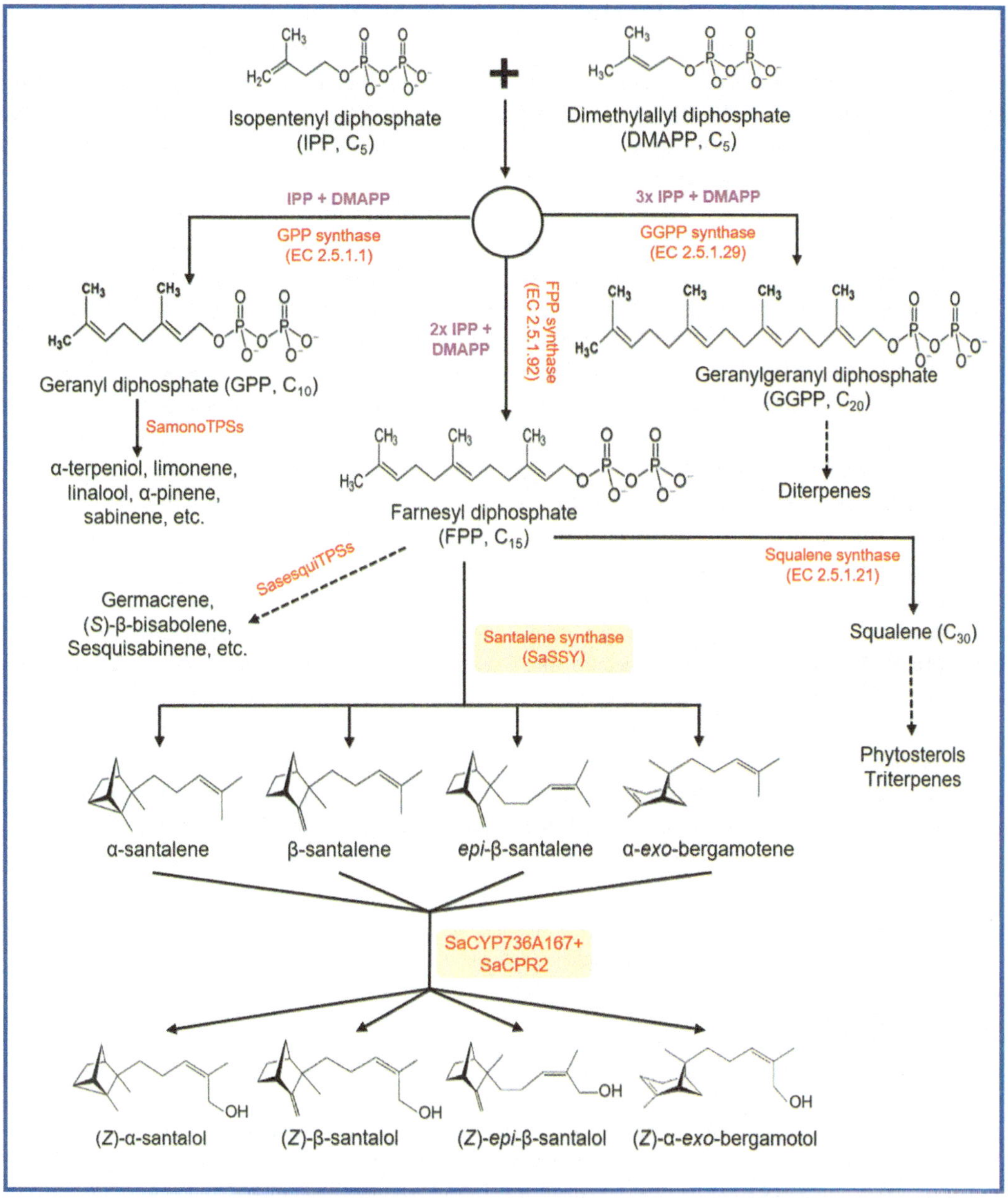

Fig. 7.2 Biosynthesis of sandalwood oil components. The IPP and DMAPP are condensed by prenyltransferases to generate C_{10} (geranyl pyrophosphate, GPP) catalyzed by GPP synthase, C_{15} (farnesyl diphosphate, FPP) catalyzed by FPP synthase, and C_{20} (geranylgeranyl diphosphate, GGPP) catalyzed by GGPP synthase. GPP and GGPP give rise to mono and diterpenes respectively, by TPS-catalyzed reactions. Squalene synthase catalyzes the condensation of two FPP molecules to produce squalene, which serves as the main precursor for phytosterols and triterpenes. Santalene synthase from *Santalum album* (SaSSY) catalyzes the formation of sesquiterpenes α-santalene, β-santalene, *epi*-β-santalene, and α-*exo*-bergamotene. These santalenes are converted into santalols by a cytochrome P450 monooxygenases (SaCYP736A167) and cytochrome P450 reductases (SaCPR2), which produce the sesquiterpene alcohols (*Z*)-α-santalol, (*Z*)-β-santalol, (*Z*)-*epi*-β-santalol, and (*Z*)-α-*exo*-bergamotol that form the fragrance-defining components of sandalwood oil. (The pathway for santalenes and santalols are adopted from Celedon et al. 2016)

Among them, SaCYP76F41, SaCYP76F42, and SaCYP76F39v1 produce both *Z*- and *E*-stereoisomers of α-santalol, β-santalol, *exo*-α-bergamotol, and *epi*-β-santalol. While SaCYP76F39v2 yielded seven products including (*Z*)-α-santalol, (*E*)-α-santalol, (*Z*)-β-santalol, (*E*)-β-santalol, (*E*)-*epi*-β-santalol, (*Z*)-*exo*-α-bergamotol, and (*E*)-*exo*-α-bergamotol, SaCYP7 6F40 only produced (*Z*)-α-santalol, (*E*)-β-santalol and (*E*)-*exo*-β-bergamotol (Diaz-Chavez et al. 2013; Rani et al. 2013). Interestingly, SaCYP76F37v1, SaCYP76F37v2, SaCYP 76F38v1, and SaCYP76F38v2 only generated *E*-stereoisomers of α-santalol, β-santalol, and *exo*-α-bergamotol. The fact that the above functionally characterized SaCYP76F enzymes produced predominantly (*E*)-stereoisomers and only small amounts of the (*Z*)-stereoisomers, which are the major components of the sandalwood oil, suggested that one or more additional enzymes are involved in sandalwood oil biosynthesis (Diaz-Chavez et al. 2013) (Table 7.1). Later, heartwood-specific transcriptome analysis in tandem with metabolite signatures in *S. album* led to the identification of a unique CYP736A167, whose transcripts were highly and selectively expressed in *S. album* heartwood. Further, in vitro and in vivo assays demonstrated that CYP736A167 was found to produce only *Z*-type products of sesquiterpene alcohols (Celedon et al. 2016) (Fig. 7.2). It was shown that SaCYP736A167 was a multisubstrate P450 enzyme that stereoselectively produces (*Z*)-α-santalol, (*Z*)-β-santalol, (*Z*)-*epi*-β-santalol, and (*Z*)-α-*exo*-bergamotol utilizing the products made by SaSSY (Fig. 7.2) (Celedon et al. 2016). Interestingly, it was shown that the 13:6:1:2 ratio of (*Z*)-α-santalol, (*Z*)-β-santalol, (*Z*)-*epi*-β-santalol, and (*Z*)-α-*exo*-bergamotol produced by SaCYP736A167 was similar to the ratio of these compounds found in commercially exploited natural sandalwood oil. It is important to note that CYPs also require, NADPH-dependant cytochrome P450 reductase (CPR), which acts as an electron donor, for its activity. Incidentally, two CPRs, SaCPR1, and SaCPR2 have been characterized in *S. album*, (Diaz-Chavez et al. 2013). It was demonstrated that the expression of SaSSY, SaCYP736A167, and SaCPR2 together in yeast, produced the four major sandalwood sesquiterpenes (*Z*)-α-santalol, (*Z*)-β-santalol, (*Z*)-*epi*-β-santalol, and (*Z*)-α-*exo*-bergamotol (Celedon et al. 2016). All the above discussed functionally characterized enzymes of *Santalum* species involved in terpene biosynthesis have been summarized in Table 7.1. Thus, the elucidation of the

Table 7.1 List of functionally characterized enzymes of *Santalum* species producing sesquiterpenes of sandalwood oil. Enzymes involved in the formation of major fragrance-defining compounds are highlighted in bold

Source	Enzyme	Products formed	References
Terpene synthases			
Santalum album	SamonoTPS1	α-terpeniol, limonene, linalool, myrcene, α-pinene, sabinene, and geraniol	(Jones et al. 2008)
Santalum album	SasesquiTPS1	Germacrene, helminthogermacrene, α-bulnesene, γ-muurolene, α- and β-selinenes	(Jones et al. 2008)
Santalum album	**SaSSY**	**α-santalene, β-santalene, *epi*-β-santalene, and α-*exo*-bergamotene**	(Jones et al. 2011)
Santalum spicatum	SspiSSY	α-santalene, β-santalene, *epi*-β-santalene, and α-*exo*-bergamotene	(Jones et al. 2011)
Santalum austroceladonicum	SausSSY	α-santalene, β-santalene, *epi*-β-santalene, and α-*exo*-bergamotene	(Jones et al. 2011)
Santalum album	SaSQS1	Sesquisabinene and sesquiphellanrene	(Srivastava et al. 2015)
Santalum album	SaSQS2	Sesquisabinene and sesquiphellanrene	(Srivastava et al. 2015)

(continued)

Table 7.1 (continued)

Source	Enzyme	Products formed	References
Santalum album	SaBS	(*S*)-β-bisabolene	(Srivastava et al. 2015)
Santalum album	SaSS	α-santalene, β-santalene, *epi*-β-santalene, *exo*-α-bergamotene, *exo*-β-bergamotene, and (*E*)-β-farnesene	(Srivastava et al. 2015)
Santalum spicatum	SspiTPS4	Sesquisabinene, β-bisabolene, γ-curcumene, β-sesquiphellandrene, α-acoradiene, β-pinene, myrcene, sabinene, α-pinene, α-terpineol, and linalool	(Moniodis et al. 2015)
Santalum album	SaTPS1	α-terpineol, sabinene, linalool, and myrcene	(Zhang et al. 2019)
Santalum album	SaTPS2	(*E*)-α-bergamotene, sesquisabinene, β-bisabolene, γ-bisabolene, 7-*epi*-sesquithujene, (*E*)-β-farnesene, α-zingiberene and linalool	(Zhang et al. 2019)
Santalum album	SaTPS3	(*E*)-β-farnesene, (*E*)-nerolidol, (*E,E*)-farnesol, γ-bisabolene, and linalool	(Zhang et al. 2019)
Santalum album	SaNES/LIS	(*E*)-nerolidol and linalool	(Zhang et al. 2019)
Cytochrome P450 enzymes			
Santalum album	Cyp76F41	(*Z*)-α-santalol, (*Z*)-α-*exo*-bergamotol, (*E*)-α-santalol, (*E*)-α-*exo*-bergamotol, (*Z*)-*epi*-β-santalol, (*Z*)-β-santalol, (*E*)-*epi*-β-santalol, (*E*)-β-santalol	(Diaz-Chavez et al. 2013)
Santalum album	Cyp76F37v1	(*E*)-α-santalol, (*E*)-α-*exo*-bergamotol, (*E*)-β-santalol	(Diaz-Chavez et al. 2013)
Santalum album	Cyp76F37v2	(*E*)-α-santalol, (*E*)-α-*exo*-bergamotol, (*E*)-β-santalol	(Diaz-Chavez et al. 2013)
Santalum album	Cyp76F42	(*Z*)-α-santalol, (*Z*)-α-*exo*-bergamotol, (*E*)-α-santalol, (*E*)-α-*exo*-bergamotol, (*Z*)-*epi*-β-santalol, (*Z*)-β-santalol, (*E*)-*epi*-β-santalol, (*E*)-β-santalol	(Diaz-Chavez et al. 2013)
Santalum album	Cyp76F39v1	(*Z*)-α-santalol, (*Z*)-α-*exo*-bergamotol, (*E*)-α-santalol, (*E*)-α-*exo*-bergamotol, (*Z*)-*epi*-β-santalol, (*Z*)-β-santalol, (*E*)-*epi*-β-santalol, (*E*)-β-santalol	(Diaz-Chavez et al. 2013)
Santalum album	Cyp76F39v2	(*Z*)-α-santalol, (*Z*)-α-*exo*-bergamotol, (*E*)-α-santalol, (*E*)-α-*exo*-bergamotol, (*Z*)-β-santalol, (*E*)-*epi*-β-santalol, (*E*)-β-santalol	(Diaz-Chavez et al. 2013)
Santalum album	Cyp76F40	(*E*)-α-*exo*-bergamotol, (*E*)-β-santalol, (*Z*)-β-santalol	(Diaz-Chavez et al. 2013)
Santalum album	Cyp76F38v1	(*E*)-α-santalol, (*E*)-α-*exo*-bergamotol, (*E*)-*epi*-β-santalol, (*E*)-β-santalol	(Diaz-Chavez et al. 2013)
Santalum album	Cyp76F38v2	(*E*)-α-santalol, (*E*)-α-*exo*-bergamotol, (*E*)-*epi*-β-santalol, (*E*)-β-santalol	(Diaz-Chavez et al. 2013)
Santalum album	**Cyp736A167**	**(*Z*)-α-santalol, (*Z*)-β-santalol, (*Z*)-*epi*-β-santalol and (*Z*)-α-*exo*-bergamotol**	(Celedon et al. 2016)
Santalum album	SaCPR1	Acts as redox partner of CYPs	(Diaz-Chavez et al. 2013)
Santalum album	**SaCPR2**	**Acts as redox partner of CYPs**	(Diaz-Chavez et al. 2013; Celedon et al. 2016)

entire santalol biosynthetic pathway in sandalwood has provided a solid foundation and opened up a bright possibility of engineering the production of commercially important santalols in alternate host systems by biotechnological approaches.

7.4 Genome of *Santalum Album*

While many transcriptomic data sets were available since 2013 (Diaz-Chavez et al. 2013; Moniodis et al. 2015; Celedon et al. 2016), it is only recently that the genome of *S. album* has been sequenced. One study used an integrated genomic, transcriptomic, and proteomic approach to assemble and annotate the genome and found that the total genome size was 221 Mb with 38,119 predicted protein-coding genes and 27.42% repetitive DNA elements. This study reported that the gene homologs for the sandalwood terpene biosynthesis characterized in earlier studies were represented in their genome. The raw sequence reads of whole-genome and RNA sequencing (SRR5150444) are deposited in the NCBI SRA database with the accession number SRP096167 (Mahesh et al. 2018). Another study attempted to produce a draft genome from *S. album* populations of Kerela, India. They found the total genome size to be 286 Mb and predicted 37,500 genes of which 30 are from the TPS gene family. In the same study, they also reconstructed the chloroplast genome of 147.25 kb. The raw sequence reads from this study are available on the NCBI SRA database with accession number SRP145575 (Dasgupta et al. 2019). In another study, the complete chloroplast genome of the *S. album* was recently sequenced and was found to be 144.1 kb in size, which contained 80 protein-coding genes, 35 *tRNA* genes, and 8 *rRNA* genes (Yang et al. 2020). Further, an in-depth understanding of the genomic data can help elucidate the molecular mechanisms involved in heartwood formation and oil biosynthesis in sandalwood trees. This information can also help implement specific strategies for increasing the production of sandalwood oil, such as genomics-assisted breeding to characterize and improve sandalwood germplasm or even genetic engineering of the sandalwood tree itself to increase oil accumulation.

7.5 Current Approaches for Sustainable Production of Sandalwood Oil

The cultivation of the sandalwood tree and the extraction of sandalwood oil is a tedious process. The tree must be at least 15 years old to possess heartwood from which the oil is distilled. The quality of the oil also varies greatly depending on the age of the tree, with 20-year-old trees giving the best quality oil, the geographical location, species and even between different trees. The demand for SO, however, is ever increasing. As described earlier, SO has various applications in the aroma and pharmaceutical industries. Hence, various attempts have been made to increase the production of this valuable product, some of which are described here.

7.5.1 Chemical Synthesis

The chemical synthesis of SO has mainly focused on creating molecules that have a similar aroma profile of naturally occurring oil. The first attempt was that of synthesizing α-santalol from camphor using a Wittig reaction, however, the product obtained was *E*-isomer whereas the naturally occurring one was *Z*-isomer (Wittig and Geissler 1953). The *Z*-isomer was finally obtained using a variant of the Wittig reaction called the Wittig-Schlosser reaction which relied upon a Claisen rearrangement, asymmetric Diels-Alder reaction, starting from norbornanone to yield a final mixture of *Z*-α-santalol and *E*-α–santalol (Corey et al. 1957; Krotz and Helmchen 1990). Chemical synthesis of β-santalene and β-santalol was achieved from camphene or camphenesultone in a six-step reaction, with an overall yield of 23% (Christenson and Willis

1980). Other efficient methods, involving only five steps, of enantioselective synthesis of Z-β-santalol from substrates like cyclopentadiene and crotonaldehyde have been developed (Fehr et al. 2009). The products produced from these methods have been noted to have a strong characteristic sandalwood odor. Ever since the study on the molecular surface of these synthesized odor molecules as a function of their biological activity, many odor relationship/structure–activity studies have given a better understanding of the fragrance-structure relationship of these molecules (Buchbauer et al. 1992, 1997). These studies eventually led to the synthesis of santalol derivatives, for example, Z-homo-β-santalol, methyl-β-santalol, and rac-Z-β-santalol. All these derivatives possess the typical sandalwood odor, however, due to multiple low-recovery steps, the yield of these compounds is no more than 20%. Currently, there are many synthetic sandalwood odorants available such as rac-orsyrol (introduced by Bush Boake Allen Ltd., now Inrandan Flavors and Fragrances Inc.), Brahmanol (Symrise), madrol (Symrise), sandranol (Symrise), ebanol (Givaudan), sandalore (Givaudan), firsantol (Firmenich), polysantol (Firmenich), and javanol (Givaudan), the first having been synthesized from dihydromyrcene and all the rest from α-campholenic aldehyde (Brocke et al. 2008; Parthasarathy and Sulochana 2016). Thus, the chemical synthesis of SO products is possible but due to the low yield, it is not viable for large-scale production.

7.5.2 Application of Elicitors

Few studies have looked into the application of elicitors to induce heartwood formation and improve the yield of oil from sandalwood trees. In a young tree, heartwood formation is induced by the injection of exogenous fungi (*Colletotrichum gloeosporioides* and *Botryosphaeria rhodian*) or pure CO_2 (99.999%) or a solution of 6-benzyladenine, 2-chloroethylphosphonic acid, copper sulfate, zinc sulphate, indole-3-butyric acid, paraquat, or ethrel into the stem (Yinglan 1994). A recent study investigated the effect of various known elicitors on heartwood formation and oil content in one and five-year-old sandalwood trees (Li et al. 2021). This study found that 6-benzyladenine was a powerful inducer of heartwood formation in young trees. Five year old sandalwood trees treated with 4 mM of 6-benzyladenine or 4 mM hydrogen peroxide increased the oil content of heartwood to 2.75% and 2.85% respectively, as compared to the control trees (0.67%). While 2 mM or 4 mM 6-benzyladenine, 2 mM methyl jasmonate, and 4 mM hydrogen peroxide significantly increased sesquiterpene content in the oil, 2 mM calcium chloride decreased sesquiterpene content as compared to trees treated with water. This study concluded that treatment of young sandalwood trees with 6-benzyladenine and hydrogen peroxide can increase cambium activity and induce the premature formation of heartwood and sandalwood oil biosynthesis (Li et al. 2021). Further work is needed to understand the cell-signaling mechanisms involved in heartwood formation but this approach can lead to more sustainable means of producing sandalwood oil by reducing the time taken for heartwood maturation.

7.5.3 Use of Tissue Culture

Callus and cell suspension cultures have been used for the production of valuable phytochemicals including various terpenoids for the past several years. These techniques have several advantages such as being the native system for production, greater control of quality, and shorter growth times. There have been several reports of generating callus, cell culture, and somatic embryos using different explants of sandalwood (Teixeira da Silva et al. 2016). However, only two studies have investigated the potential of tissue culture for producing terpenoids. It was reported that green and dense calli generated from young hypocotyl segments obtained from aseptically germinated seeds of *S. album* were able to produce a range of terpene molecules, including santalenes and santalols. This indicated the ability of in vitro tissues to produce terpenes of the sandalwood oil profile. However, the sesquiterpene profiles of calli differed from the commercial oil profile in the relative quantities of the desired molecules (Crovadore et al. 2012). In another recent study, highly proliferating cell lines of sandalwood were generated using seed explants by medium optimization for the production of squalene. Using a 7 L bioreactor with optimized parameters 5.5 mg of squalene (constituting about 27% of total terpenes) was obtained per gram dry weight on the 28th day. In addition to squalene other sesquiterpenes such as nerolidol (1%), farnesol (0.74%), and diterpenes like geranyllinalool (1.45%), geranylgeraniol (65.86%), geranylgeranyl acetate (diterpene; 3.56%) were produced in the cell cultures. In contrast to the previous study, here there was no production of santalenes and santalols (Rani et al. 2018). These two above studies suggest that sandalwood cell cultures have an active terpenoid pathway to synthesize a variety of terpenes and by manipulating the media conditions, it is possible to achieve desired metabolite production. Very recently, electro-elicitation of calli and cell suspensions in *Gyrinops walla*, which is known to produce agarwood rich in fragrant terpenes led to the production of sesquiterpenes γ-selinene, β-caryophyllene, α-cadinol, and α-guaiene in both calli and cell suspension (Munasinghe et al. 2021). A similar approach in sandalwood cultures could be explored for the induction of santalols in heartwood-derived callus or suspension cultures. Further, as pathway has been elucidated and protocols for genetic transformation have been reported in sandalwood (Teixeira da Silva et al. 2016), transgenic culture or trees overexpressing the genes of the biosynthetic pathway under strong promoters could be explored for the production of santalols.

7.5.4 Production of Sandalwood Terpenes in Alternate Host Systems: *Escherichia Coli*

E. coli is a host system commonly used in the lab for the production of valuable compounds including isoprenoids (Ward et al. 2018). Due to its inexpensive culture conditions, rapid growth, and ease of genetic modifications, it is the preferred host in most biotechnological applications (Pontrelli et al. 2018). *E. coli* possesses the MEP pathway that forms the precursors IPP and DMAPP for the production of terpenes (Kuzuyama and Seto 2012; Wang et al. 2020). Production of α-santalene was reported in *E. coli* by the expression of santalene synthase gene (SanSyn) from *Clausena lansium* (Schalk 2011). In this case, the *E. coli* strain was grown in a medium supplemented with 2 mg/ml mevalonate and achieved α-santalene production of 250 mg/L in 48 h. However, this host lacks certain post-translational modification machinery necessary for the expression of functional eukaryotic plant enzymes. Also, this host has no cell membrane systems, as cytochrome P450 enzymes are usually anchored to the endoplasmic reticulum (ER) and there is no CPR system present in *E. coli* for mediating electron transfer. Hence, most CYPs are not functional when expressed in this host. This problem has been partially solved by engineering a chimeric CYP with a membrane-binding domain that anchors the enzyme to the cell membrane, mimicking the conditions inside a plant ER. This approach has limited success with very few examples of functional expression of CYPs leading to the production of terpenoids in *E. coli* (Chang et al. 2007; Wilson and Roberts 2012; Hausjell et al. 2018).

7.5.5 Production of Sandalwood Terpenes in Alternate Host Systems: *Saccharomyces Cerevisiae*

S. cerevisiae is the preferred host system for the production of plant-based natural compounds and especially sesquiterpenes (Gohil et al. 2017; Rahmat and Kang 2020). Favorable qualities of this host system include, single-celled nature, rapid growth, inexpensive culture conditions, tolerance to abiotic/industrial stress conditions, ease of large scale production, ease of genetic engineering due to decades of research and development, presence of cell organelles, post-translational modification i.e., being able to functionally express most CYP enzyme systems.

S. cerevisiae can synthesize IPP and DMAPP through its MVA pathway (Kuzuyama and Seto 2012). This pathway is mainly responsible for the biosynthesis of ergosterol, the genes involved in this pathway have been well characterized and is the main focus when engineering for terpene production (Jordá and Puig 2020). In this pathway, *ERG10* that encodes acetoacetyl-CoA thiolase catalyzes the condensation of two molecules of acetyl-CoA to acetoacetyl-CoA which is then converted into 3-hydroxy-3-methyl-glurary1-CoA (HMG-CoA) by ERG13 (HMG- CoA synthase). HMG-CoA is then reduced by HMG1 or HMG2 (HMG-CoA reductase) to produce mevalonic acid. This reaction is the key rate-limiting step in the pathway after which IPP is yielded through a series of conversions successively, catalyzed by ERG12 (mevalonate-5-kinase), ERG8 (phosphomevalonate kinase), and ERG19 (mevalonate pyrophosphate decarboxylase). Finally, the reversible conversion between IPP and DMAPP is catalyzed by IDI1 (isopentyl pyrophosphate isomerase). DMAPP and IPP are then condensed to form FPP by the action of ERG20 (FPP synthase). To increase IPP and DMAPP precursor production, a lot of effort has been put into optimizing the MVA pathway in *S. cerevisiae*. A key rate-limiting reaction is catalyzed by the enzyme HMG1 and HMG2, both these enzymes possess a trans-membrane domain which anchors it to the ER as well as a catalytic domain (Donald et al. 1997). The overexpression of a truncated form of HMG1 (tHMG1, only catalytic domain of HMG1) has been shown to greatly improve the production of terpenes in *S. cerevisiae* (Dai et al. 2013; Huang et al. 2019). UPC2 is a transcription factor that plays a key role in the regulation of the MVA pathway. The overexpression of its mutant form, UPC2-1 has been shown to greatly increase expression of the MVA pathway and the concentration of the precursors (Ro et al. 2006). Similarly, it has also been shown that the MVA pathway can be significantly upregulated by manipulation of the phospholipid biosynthesis pathway (Ranganathan et al. 2020). In *S. cerevisiae* cells, there is a high demand for sterols caused by their rapid growth and division. To meet this demand FPP, the direct precursor to sesquiterpenes is rapidly consumed by squalene synthase (ERG9) to produce squalene. Then through a series of cyclization and modification reactions, squalene is converted to ergosterol (Daum et al. 1998). To prevent the depletion of the FPP pool, strategies involving the downregulation of *ERG9* by replacing the native *ERG9* promoter with a repressible *MET3* promoter have been very successful in increasing the production of sesquiterpenes (Ro et al. 2006; Asadollahi et al. 2008). The other branch pathway that consumes FPP is by the action of LPP1 and DPP1, both of which encode lipid phosphate phosphatases, forms farnesol (Toke et al. 1998a, b). One of the first attempts of producing sandalwood sesquiterpenes in *S. cerevisiae* was by expressing SanSyn, and multiple copies of tHMG1 together with downregulation of *ERG9* (Scalcinati et al. 2012). In the same study, it was shown that the knockout of *LPP1* and *DPP1* was shown to result in a significant improvement of santalene production (up to 91.96 mg/L) and inhibition of farnesol generation (Scalcinati et al. 2012). Another key metabolite in the MVA pathway is acetyl-CoA. Acetyl-CoA is massively consumed in the peroxisome and cytosol by the glyoxylate cycle in which CIT2 (peroxisomal citrate synthase) and MLS1 (cytosolic malate) catalyze the synthesis of citrate and malate respectively from acetyl-CoA (Chen et al. 2012). Knockout of these two genes have been reported to improve α-santalene production in *S. cerevisiae* by four-fold, at ~40 mg/L (Chen et al. 2013). NADPH is an important cofactor that participates in redox reactions and is essential for the enzymatic reactions catalyzed by tHMG and CYPs, hence many studies have focused on improving the concentration of NADPH within the cell to improve terpenoid production (Celton et al. 2012; Li and Zhang 2015; Paramasivan and Mutturi 2017). Scalcinati et. al. (2012) tried to improve NADPH concentration by deleting

GDH1 (NADP- dependent glutamate dehydrogenase) and overexpressing GDH2 (NAD-dependent glutamate dehydrogenase), which resulted in a decreased consumption of NADPH by the ammonium assimilation pathway and in turn led to a significant increase in α-santalene yield (0.036 Cmmol (g biomass)$^{-1}$ h^{-1}) in engineered *S. cerevisiae* (Scalcinati et al. 2012).

Another key strategy used in *S. cerevisiae* to improve expression is the use of galactose inducible promoter (GAL promoter). The GAL regulatory system is very well understood in this host and can be easily induced or repressed by the addition or removal of galactose from the growth medium (Ro et al. 2006; Paddon et al. 2013). In a recent study, key enzymes for sandalwood sesquiterpene production tHMG1, UPC2-1, SaSSY, CYP736A167, and SaCPR2 were expressed under the influence of strong GAL promoters. Meanwhile, the transcriptional activator of GAL promoter, GAL4, was overexpressed to increase the inducible effect of galactose, and PGM2 (phosphoglucomutase) was also overexpressed to increase the uptake of galactose. These strategies resulted in an *S. cerevisiae* strain that produced 1.2 g/L of Z-α-santalol in a two-phase fed-batch fermentation reaction after 84 h. The same study reported the native strain of *S. cerevisiae* (BY4741) to produce only 1.3 g/L of santalols when SaSSY and CYP736A167 were expressed (Zha et al. 2020).

7.5.6 Production of Oil in Other Host Systems

Other, lesser established host systems have been explored for the heterologous production of santalenes. These include *Yarrowia lipolytica, Nicotiana tobacum,* and *Physcomitrella patens* (Zhan et al. 2014; Yin and Wong 2019; Jia et al. 2019). *Y. lipolytica* is an oleaginous yeast known for its metabolic prowess. This organism is recently gaining popularity in the field of synthetic biology due to its high tolerance for metabolic stress and its ability to utilize a wide variety of raw carbon sources (e.g., molasses), including hydrophobic ones like palm oil (Tai 2012; Louhasakul et al. 2016; Do et al. 2019; Larroude et al. 2019). Overexpression of upstream MVA pathway genes and modulation of carbon source concentration in the medium resulted in 27.92 mg/L α-santalene production (Jia et al. 2019). Further improvement in production is still a challenge due to the limited genetic manipulation tools available for *Y. lipolytica*. In heterologous plants, so far only santalene production has been explored in moss (*P. patens*) and tobacco (*N. tobacum*). In moss, targeting of sandalwood santalene synthase to the plastids yielded 0.039 mg/g dry weight of α/β santalene, whereas coexpression of cytosolic santalene synthase and *HMGR* resulted in 0.022 mg/g dry weight (Zhan et al. 2014). Transient and stable overexpression of SaSSY in combination with HMGR and FPPS in cytosol or plastids of tobacco showed production of α-santalene, α-bergamotene, *Epi-β*-santalene, and of -santalene in nanogram quantities (Yin and Wong 2019). However, in both the studies the yield of santalenes was very low compared to microbial hosts and further optimization in engineering is needed to achieve higher yields in plants. Chloroplast engineering or compartmentalized metabolic engineering of the santalol pathway may be explored for achieving higher yields in tobacco.

7.6 Conclusion

Sandalwood is one of the most prized commodities in the fragrance industry. Its warm, spicy, woody scent makes it a key ingredient in many perfume products. Sandalwood has been used since ancient times and has a rich cultural history with India. However, overexploitation of natural populations, habitat loss, and poaching combined with the slow growth of the tree and limited land resources have rendered conventional cultivation of the tree unsustainable and unable to meet the commercial demand. This has prompted the investigation into other sources of sandalwood oil. Analysis of the oil showed the active compound to be α-santalol and β-santalol. These are sesquiterpene alcohols that are abundantly found in the oil. Chemical synthesis of the correct isomers of these compounds that have a similar aroma profile to their natural counterparts have been achieved since 1990, however, low yield rates, multiple reaction steps, and the release of environmental pollutants prevent these approaches from reaching industrial scale. The current understanding of biosynthetic pathways for these compounds is well established. This has led to the engineering of the metabolic pathway in alternate host systems for the production of sandalwood oil. The most significant progress in this area is made in the production of santalols in host organisms, especially *S. cerevisiae,* by the heterologous expression of sandalwood genes. In the best study, 1.3 g/L of santalols was produced by engineered *S. cerevisiae* in a fed-batch fermentation system. While this achievement is remarkable, high input cost, low yield, and current market conditions make this approach economically unviable. There is an urgent need for further improvements in the yield of sandalwood oil. This may be achieved through a more combined approach of all the studies discussed above. It may be possible to increase sesquiterpene yield in *S. cerevisiae* by further diverting the carbon flux from other competing pathways to the MVA pathway through genetic engineering. Alternate substrates, such as ethanol, can also be explored to increase acetyl-CoA flux. Genetic modification of the *Santalum* tree or cell cultures itself remains largely unexplored and may be further investigated for optimization of SO production. Also, heterologous engineering in high-biomass producing plants such as tobacco could be further explored. If successful, these approaches could be a viable option for the sustainable production of sandalwood oil. In conclusion, while many large strides have been taken in the understanding of sandalwood oil biosynthesis and its production in alternative hosts, current technologies and approaches have not yet achieved the economic viability required to move away from the conventional plantation and sustainable production. Further research is required to improve the yield and lower the input cost and the impact on the environment for sustainable production of sandalwood oil.

Acknowledgements This work was supported by the GAP-454 project (BIRAC funded project BIRAC/BT/SB0066/02/19) of CSIR-Central Institute of Medicinal and Aromatic Plants. The institutional communication number for this article is CIMAP/PUB/2021/June/27.

References

Akhtar TA, Matsuba Y, Schauvinhold I et al (2013) The tomato cis-prenyltransferase gene family. Plant J 73:640–652. https://doi.org/10.1111/tpj.12063

Asadollahi MA, Maury J, Møller K et al (2008) Production of plant sesquiterpenes in Saccharomyces cerevisiae: effect of ERG9 repression on sesquiterpene biosynthesis. Biotechnol Bioeng 99:666–677. https://doi.org/10.1002/bit.21581

Baldovini N, Delasalle C, Joulain D (2011) Phytochemistry of the heartwood from fragrant Santalum species: a review. Flavour Fragr J 26:7–26

Bohlmann J, Meyer-Gauen G, Croteau R (1998) Plant terpenoid synthases: molecular biology and phylogenetic analysis. Proc Natl Acad Sci USA. https://doi.org/10.1073/pnas.95.8.4126

Brand JE, Ryan PC, Williams MR (1999) Establishment and growth of sandalwood (antalum spicatum) in south-western Australia: the Northampton pilot trial. Aust for 62:33–37. https://doi.org/10.1080/00049158.1999.10674760

Brocke C, Eh M, Finke A (2008) Recent developments in the chemistry of sandalwood odorants. Chem Biodivers. https://doi.org/10.1002/cbdv.200890080

Buchbauer G, Winiwarter S, Wolschann P (1992) Surface comparisons of some odour molecules: conformational calculations on sandalwood odour V. J Comput Aided Mol Des. https://doi.org/10.1007/BF00126216

Buchbauer G, Spreitzer H, Koller U (1997) (Z)-homo-β-Santalol: synthese und Geruch. Monatshefte Fur Chemie 128:619–624. https://doi.org/10.1007/BF00807592

Burdock GA, Carabin IG (2008) Safety assessment of sandalwood oil (*Santalum album* L.). Food Chem Toxicol 46:421–432

Celedon JM, Chiang A, Yuen MMS et al (2016) Heartwood-specific transcriptome and metabolite signatures of tropical sandalwood (*Santalum album*) reveal the final step of (Z)-santalol fragrance biosynthesis. Plant J 86:289–299. https://doi.org/10.1111/tpj.13162

Celton M, Goelzer A, Camarasa C et al (2012) A constraint-based model analysis of the metabolic consequences of increased NADPH oxidation in Saccharomyces cerevisiae. Metab Eng 14:366–379. https://doi.org/10.1016/j.ymben.2012.03.008

Chang MCY, Eachus RA, Trieu W et al (2007) Engineering Escherichia coli for production of functionalized terpenoids using plant P450s. Nat Chem Biol 3:274–277. https://doi.org/10.1038/nchembio875

Chen F, Tholl D, Bohlmann J, Pichersky E (2011) The family of terpene synthases in plants: a mid-size family of genes for specialized metabolism that is highly diversified throughout the kingdom. Plant J. https://doi.org/10.1111/j.1365-313X.2011.04520.x

Chen Y, Daviet L, Schalk M et al (2013) Establishing a platform cell factory through engineering of yeast acetyl-CoA metabolism. Metab Eng 15:48–54. https://doi.org/10.1016/j.ymben.2012.11.002

Chen Y, Siewers V, Nielsen J (2012) Profiling of cytosolic and peroxisomal acetyl-CoA metabolism in Saccharomyces cerevisiae. PLoS One 7. https://doi.org/10.1371/journal.pone.0042475

Christenson PA, Willis BJ (1980) East Indian Sandalwood Oil. 2. Stereoselective synthesis of (±)-epi-β-santalene and (±)-epi-β-santalol. J Org Chem 45:3068–3072. https://doi.org/10.1021/jo01303a026

Corey EJ, Chow SW, Scherrer RA (1957) The synthesis of -Santalene and of trans-Δ-Iso-a-santalene. J Am Chem Soc 79:5773–5777. https://doi.org/10.1021/ja01578a049

Crovadore J, Schalk M, Lefort F (2012) Selection and Mass Production of Santalumï¿¿Album L. Calli for induction of sesquiterpenes. Biotechnol Biotechnol Equip 26:2870–2874. https://doi.org/10.5504/BBEQ.2012.0028

Dai Z, Liu Y, Zhang X et al (2013) Metabolic engineering of Saccharomyces cerevisiae for production of ginsenosides. Metab Eng 20:146–156. https://doi.org/10.1016/j.ymben.2013.10.004

Dasgupta MG, Ulaganathan K, Dev SA, Balakrishnan S (2019) Draft genome of *Santalum album* L. provides genomic resources for accelerated trait improvement. Tree Genet Genomes 15:1–15. https://doi.org/10.1007/s11295-019-1334-9

Daum G, Lees ND, Bard M, Dickson R (1998) Biochemistry, cell biology and molecular biology of lipids of Saccharomyces cerevisiae. Yeast

De Groot AC, Schmidt E (2016) Essential oils, part I: introduction. Dermatitis 27:39–42

De Groot AC, Schmidt E (2017) Essential oils, part VI: sandalwood oil, ylang-ylang oil, and jasmine absolute. Dermatitis 28:14–21

Demissie ZA, Erland LAE, Rheault MR, Mahmoud SS (2013) The biosynthetic origin of irregular monoterpenes in lavandula: isolation and biochemical characterization of a novel cis-prenyl diphosphate synthase gene, lavandulyl diphosphate synthase. J Biol Chem. https://doi.org/10.1074/jbc.M112.431171

Dhanya B, Viswanath S, Purushothman S (2010) Sandal (*santalum album* L.) conservation in Southern India: A review of policies and their impacts. J Trop Agric

Di Girolamo A, Durairaj J, van Houwelingen A et al (2020) The santalene synthase from cinnamomum camphora: reconstruction of a sesquiterpene synthase from a monoterpene synthase. Arch Biochem Biophys. https://doi.org/10.1016/j.abb.2020.108647

Diaz-Chavez ML, Moniodis J, Madilao LL, et al (2013) Biosynthesis of Sandalwood Oil: Santalum album CYP76F Cytochromes P450 produce Santalols and Bergamotol. PLoS One 8. https://doi.org/10.1371/journal.pone.0075053

Do DTH, Theron CW, Fickers P (2019) Organic wastes as feedstocks for non-conventional yeast-based bioprocesses. Microorganisms

Donald KAG, Hampton RY, Fritz IB (1997) Effects of overproduction of the catalytic domain of 3-hydroxy-3- methylglutaryl coenzyme a reductase on squalene synthesis in Saccharomyces cerevisiae. Appl Environ Microbiol 63:3341–3344. https://doi.org/10.1128/aem.63.9.3341-3344.1997

Eisenreich W, Bacher A, Arigoni D, Rohdich F (2004) Biosynthesis of isoprenoids via the non-mevalonate pathway. Cell Mol Life Sci

Fehr C, Magpantay I, Arpagaus J et al (2009) Enantioselective synthesis of (-)-β-Santalol by a copper-catalyzed enynol cyclization-fragmentation reaction. Angew Chemie Int Ed 48:7221–7223. https://doi.org/10.1002/anie.200903449

Gohil N, Panchasara H, Patel S et al (2017) Book review: recent advances in yeast metabolic engineering. Front Bioeng Biotechnol 5:71. https://doi.org/10.3389/fbioe.2017.00071

Gutensohn M, Nagegowda DA, Dudareva N (2013) Involvement of compartmentalization in monoterpene and sesquiterpene biosynthesis in plants. In: Bach TJ, Rohmer M (eds) Isoprenoid synthesis in plants and microorganisms: new concepts and experimental approaches. Springer, NY, pp 155–169

Hausjell J, Halbwirth H, Spadiut O (2018) Recombinant production of eukaryotic cytochrome P450s in microbial cell factories. Biosci Rep 38:20171290

Huang J, Zha W, An T et al (2019) Identification of RoCYP01 (CYP716A155) enables construction of engineered yeast for high-yield production of betulinic acid. Appl Microbiol Biotechnol 103:7029–7039. https://doi.org/10.1007/s00253-019-10004-z

Jia D, Xu S, Sun J et al (2019) Yarrowia lipolytica construction for heterologous synthesis of α-santalene and fermentation optimization. Appl Microbiol Biotechnol 103:3511–3520. https://doi.org/10.1007/s00253-019-09735-w

Jia Q, Chen F (2016) Catalytic functions of the isoprenyl diphosphate synthase superfamily in plants: a growing repertoire. Mol Plant

Jones CG, Moniodis J, Zulak KG et al (2011) Sandalwood fragrance biosynthesis involves sesquiterpene synthases of both the terpene synthase (TPS)-a and TPS-b subfamilies, including santalene synthases. J Biol Chem 286:17445–17454. https://doi.org/10.1074/jbc.M111.231787

Jones CG, Keeling CI, Ghisalberti EL, et al (2008) Isolation of cDNAs and functional characterisation of two multi-product terpene synthase enzymes from sandalwood, *Santalum album* L. Arch Biochem Biophys. https://doi.org/10.1016/j.abb.2008.05.008

Jordá T, Puig S (2020) Regulation of ergosterol biosynthesis in saccharomyces cerevisiae. Genes (basel). 11:1–18

Kellogg BA, Poulter CD (1997) Chain elongation in the isoprenoid biosynthetic pathway. Curr Opin Chem Biol 1:570–578. https://doi.org/10.1016/S1367-5931(97)80054-3

Krotz A, Helmchen G (1990) Total syntheses of sandalwood fragrances: (Z)- and (E)-β-santalol and their enantiomers, ent-β-santalene. Tetrahedron Asymmetry 1:537–540. https://doi.org/10.1016/S0957-4166(00)80544-3

Kuzuyama T, Seto H (2012) Two distinct pathways for essential metabolic precursors for isoprenoid biosynthesis. Proc Japan Acad Ser B Phys Biol Sci

Larroude M, Park YK, Soudier P et al (2019) A modular golden gate toolkit for Yarrowia lipolytica synthetic biology. Microb Biotechnol. https://doi.org/10.1111/1751-7915.13427

Li Y, Zhang X, Cheng Q et al. (2021) Elicitors modulate young sandalwood (*Santalum album* l.) growth, heartwood formation, and concrete oil synthesis. Plants. https://doi.org/10.3390/plants10020339

Li J, Zhang Y (2015) Modulating betulinic acid production in Saccharomyces cerevisiae by managing the intracellular supplies of the co-factor NADPH and oxygen. J Biosci Bioeng 119:77–81. https://doi.org/10.1016/j.jbiosc.2014.06.013

Liao P, Wang H, Hemmerlin A et al (2014) Past achievements, current status and future perspectives of studies on 3-hydroxy-3-methylglutaryl-CoA synthase (HMGS) in the mevalonate (MVA) pathway. Plant Cell Rep 33:1005–1022

Lombard J, Moreira D (2011) Origins and early evolution of the mevalonate pathway of isoprenoid biosynthesis in the three domains of life. Mol Biol Evol. https://doi.org/10.1093/molbev/msq177

Louhasakul Y, Cheirsilp B, Prasertsan P (2016) Valorization of palm oil mill effluent into lipid and cell-bound lipase by marine yeast Yarrowia lipolytica and their application in biodiesel production. Waste Biomass Valoriz 7:417–426. https://doi.org/10.1007/s12649-015-9451-7

Mahesh HB, Subb P, Advani J et al (2018) Multi-omics driven assembly and annotation of the sandalwood (*Santalum album*) genome. Plant Physiol 176:2772–2788. https://doi.org/10.1104/pp.17.01764

Miettinen K, Pollier J, Buyst D et al (2017) The ancient CYP716 family is a major contributor to the diversification of eudicot triterpenoid biosynthesis. Nat Commun. https://doi.org/10.1038/ncomms14153

Moniodis J, Jones CG, Barbour EL et al (2015) The transcriptome of sesquiterpenoid biosynthesis in heartwood xylem of Western Australian sandalwood (Santalum spicatum). Phytochemistry. https://doi.org/10.1016/j.phytochem.2014.12.009

Moy RL, Levenson C (2017) Sandalwood album oil as a botanical therapeutic in dermatology. J Clin Aesthet Dermatol

Munasinghe DSP, Somaratne S, Weerakoon SR, Ranasinghe C (2021) In vitro production of sesquiterpenes by electro-elicitation in Gyrinops walla Gaetner calli and cell suspension cultures. Eurobiotech J 5:70–78. https://doi.org/10.2478/ebtj-2021-0011

Muratore A, Clinet JC, Duñach E (2010) Synthesis of new exo- and endo-3,8-dihydro-β-santalols and other norbornyl-derived alcohols. Chem Biodivers. https://doi.org/10.1002/cbdv.200900119

Nagegowda DA (2010) Plant volatile terpenoid metabolism: biosynthetic genes, transcriptional regulation and subcellular compartmentation. FEBS Lett 584:2965–2973

Nagegowda DA, Dudareva N (2008) Plant biochemistry and biotechnology of flavor compounds and essential oils. Medicinal plant biotechnology: from basic research to industrial applications. Wiley-VCH Verlag GmbH, pp 469–492

Nagegowda DA, Gupta P (2020) Advances in biosynthesis, regulation, and metabolic engineering of plant specialized terpenoids. Plant Sci 294

Paddon CJ, Westfall PJ, Pitera DJ et al (2013) High-level semi-synthetic production of the potent antimalarial artemisinin. Nature 496:528–532. https://doi.org/10.1038/nature12051

Paramasivan K, Mutturi S (2017) Regeneration of NADPH Coupled with HMG-CoA reductase activity increases squalene synthesis in Saccharomyces cerevisiae. J Agric Food Chem. https://doi.org/10.1021/acs.jafc.7b02945

Parthasarathy R, Sulochana N (2016) Simple synthesis of sandalwood odorant rac-Osyrol® and its ethoxy homologue. Flavour Fragr J. https://doi.org/10.1002/ffj.3287

Pontrelli S, Chiu TY, Lan EI et al (2018) Escherichia coli as a host for metabolic engineering. Metab Eng 50:16–46

Rahmat E, Kang Y (2020) Yeast metabolic engineering for the production of pharmaceutically important secondary metabolites. Appl Microbiol Biotechnol 104:4659–4674

Ranganathan PR, Nawada N, Narayanan AK, Rao DKV (2020) Triglyceride deficiency and diacylglycerol kinase1 activity lead to the upregulation of mevalonate pathway in yeast: A study for the development of potential yeast platform for improved production of triterpenoid. Biochim Biophys Acta - Mol Cell Biol Lipids. https://doi.org/10.1016/j.bbalip.2020.158661

Rani A, Ravikumar P, Reddy MD, Kush A (2013) Molecular regulation of santalol biosynthesis in *Santalum album* L. Gene 527:642–648. https://doi.org/10.1016/j.gene.2013.06.080

Rani A, Meghana R, Kush A (2018) Squalene production in the cell suspension cultures of Indian sandalwood (*Santalum album* L.) in shake flasks and air lift bioreactor. Plant Cell Tissue Organ Cult 135:155–167. https://doi.org/10.1007/s11240-018-1452-3

Rivera SB, Swedlund BD, King GJ et al (2001) Chrysanthemyl diphosphate synthase: Isolation of the gene and characterization of the recombinant non-head-to-tail monoterpene synthase from Chrysanthemum cinerariaefolium. Proc Natl Acad Sci U S A. https://doi.org/10.1073/pnas.071543598

Ro DK, Paradise EM, Quellet M et al (2006) Production of the antimalarial drug precursor artemisinic acid in engineered yeast. Nature. https://doi.org/10.1038/nature04640
Rohdich F, Kis K, Bacher A, Eisenreich W (2001) The non-mevalonate pathway of isoprenoids: genes, enzymes and intermediates. Curr Opin Chem Biol 5 (5):535–540. https://doi.org/10.1016/S1367-5931(00)00240-4
Sallaud C, Rontein D, Onillon S et al (2009) A novel pathway for sesquiterpene biosynthesis from Z, Z-farnesyl pyrophosphate in the wild tomato Solanum habrochaites. Plant Cell 21:301–317. https://doi.org/10.1105/tpc.107.057885
Sandeep C, Manohara TN (2019) Sandalwood in India: historical and cultural significance of *Santalum album* L. as a basis for its conservation. NeBIO 10:235–241
Scalcinati G, Knuf C, Partow S et al (2012) Dynamic control of gene expression in Saccharomyces cerevisiae engineered for the production of plant sesquitepene α-santalene in a fed-batch mode. Metab Eng 14:91–103. https://doi.org/10.1016/j.ymben.2012.01.007
Schalk M (2011) Method for producing alpha santalene. 39
Seemann M, Tse Sum Bui B, Wolff M et al (2006) Isoprenoid biosynthesis in plant chloroplasts via the MEP pathway: direct thylakoid/ferredoxin-dependent photoreduction of GcpE/IspG. FEBS Lett 580:1547–1552. https://doi.org/10.1016/j.febslet.2006.01.082
Sindhu R, Arora S (2013) Sandalwood oil: phytochemical and pharmacological updates, pp 181–191
Srivastava PL, Daramwar PP, Krithika R et al (2015) Functional characterization of novel sesquiterpene synthases from Indian Sandalwood, Santalum album. Sci Rep https://doi.org/10.1038/srep10095
Subasinghe SMCUP (2013) Sandalwood research: a global perspective. J Trop For Environ. https://doi.org/10.31357/jtfe.v3i1.1117
Tai M (2012) Metabolic engineering of oleaginous yeast for the production of biofuels. Diss—MIT
Teixeira da Silva JA, Kher MM, Soner D et al (2016) Sandalwood: basic biology, tissue culture, and genetic transformation. Planta 243:847–887
Tholl D (2015) Biosynthesis and biological functions of terpenoids in plants. Adv Biochem Eng Biotechnol 148:63–106. https://doi.org/10.1007/10_2014_295
Thomson LAJ (2020) Looking ahead–global sandalwood production and markets in 2040, and implications for pacific Island producers. Aust for 83:245–254. https://doi.org/10.1080/00049158.2020.1841441
Toke DA, Bennett WL, Dillon DA et al (1998) Isolation and characterization of the Saccharomyces cerevisiae DPP1 gene encoding diacylglycerol pyrophosphate phosphatase. J Biol Chem 273:3278–3284. https://doi.org/10.1074/jbc.273.6.3278
Toke DA, Bennett WL, Oshiro J et al (1998) Isolation and characterization of the Saccharomyces cerevisiae LPP1 gene encoding a Mg2+-independent phosphatidate phosphatase. J Biol Chem 273:14331–14338. https://doi.org/10.1074/jbc.273.23.14331
Vranová E, Coman D, Gruissem W (2013) Network analysis of the MVA and MEP pathways for isoprenoid synthesis. Annu Rev Plant Biol 64:665–700
Wang Q, Quan S, Xiao H (2019) Towards efficient terpenoid biosynthesis: manipulating IPP and DMAPP supply. Bioresour Bioprocess 6
Wang Z, Sun J, Yang Q, Yang J (2020) Metabolic engineering escherichia coli for the production of lycopene. Molecules 25
Ward VCA, Chatzivasileiou AO, Stephanopoulos G (2018) Metabolic engineering of Escherichia coli for the production of isoprenoids. FEMS Microbiol Lett 365:79
Wilson SA, Roberts SC (2012) Recent advances towards development and commercialization of plant cell culture processes for the synthesis of biomolecules. Plant Biotechnol J 10:249–268
Wittig G, Geissler G (1953) Zur Reaktionsweise des Pentaphenyl-phosphors und einiger Derivate. Justus Liebigs Ann Chem 580:44–57. https://doi.org/10.1002/jlac.19535800107
Yang D, Qiu Q, Xu L et al (2020) The complete chloroplast genome sequence of *Santalum album*. Mitochondrial DNA Part B Resour 5:406–407. https://doi.org/10.1080/23802359.2019.1704199
Yin JL, Wong WS (2019) Production of santalenes and bergamotene in Nicotiana tabacum plants. PLoS One 14. https://doi.org/10.1371/journal.pone.0203249
Yinglan L (1994) Undefined studies on forcing heartwood formation in santalum album. en.cnki.com.cn
Zerbe P, Bohlmann J (2015) Plant diterpene synthases: exploring modularity and metabolic diversity for bioengineering. Trends Biotechnol
Zha W, An T, Li T et al (2020) Reconstruction of the biosynthetic pathway of santalols under control of the gal regulatory system in yeast. ACS Synth Biol 9:449–456. https://doi.org/10.1021/acssynbio.9b00479
Zhan X, Zhang YH, Chen DF, Simonsen HT (2014) Metabolic engineering of the moss physcomitrella patens to produce the sesquiterpenoids patchoulol and α/β-santalene. Front Plant Sci 5. https://doi.org/10.3389/fpls.2014.00636
Zhang X, Niu M, Teixeira Da Silva JA et al (2019) Identification and functional characterization of three new terpene synthase genes involved in chemical defense and abiotic stresses in *Santalum album*. BMC Plant Biol. https://doi.org/10.1186/s12870-019-1720-3
Zhang X, Teixeira da Silva JA, Niu M et al (2021) Functional characterization of an Indian sandalwood (*Santalum album* L.) dual-localized bifunctional nerolidol/linalool synthase gene involved in stress response. Phytochemistry. https://doi.org/10.1016/j.phytochem.2020.112610
Zhao L, Chang WC, Xiao Y et al (2013) Methylerythritol phosphate pathway of isoprenoid biosynthesis. Annu Rev Biochem 82:497–530

Sandalwood Essential Oil and Its Phyto-Chemistry

8

N. R. Raghavendra and H. B. Mahesh

Abstract

Indian sandalwood (*Santalum album* L.) is an exquisite evergreen woody tropical tree which yields one of the world's most expensive essential oil from its matured heartwood (~US$2600 l^{-1}). The tree accumulates sesquiterpenes, second important group of active compounds in essential oil of plants. α- and β-santalols ($C_{15}H_{24}O$) in addition to α-and β-santalenes are the major constituents. The minor constituents of essential oil includes lanceol, nuciferol, bisabolol and the sesquiterpene hydrocarbons such as, bergamotenes, α-, β- and γ-curcumenes, β-bisabolene. Oil yield vary from 3 to 4% through steam distillation (48–72 h). Sandalwood oil having incredible signature in their base and top note, Cis-α-santalol is responsible for the light woody smell, and Cis-β-santalol is more linked to the stronger woody smell with a distinctive mark which fetch high value in the perfume and fragrance industries. In addition to this, sandalwood oil is well known to aid wide range of ailments and it has long antiquity of pharmacology and aromatherapy.

N. R. Raghavendra (✉) · H. B. Mahesh
Department of Genetics and Plant Breeding, College of Agriculture, Mandya, University Agricultural Sciences, Bangalore, India

8.1 Introduction

The genus *Santalum* belongs to the family *Santalaceae* and it consists of 19 species viz., *S. accuminatum, S. album, S. austrocaledonicum, S. boninese, S. ellipticum, S. fernandezianum, S. freycinetianum, S. haleakalae, S. insulare, S. involutum, S. lanceolatum, S. macgregorii, S. murrayanum, S. obtusifolium, S. peniculatum, S. pyrularium, S. salicifolium, S. spicatum* and *S. yasi* occurring mainly in Australia and the Pacific (Subasinghe 2013). The *Santalum album* L. (East Indian Sandalwood) is highly prized tree of global fame (Mahesh et al. 2018) and distribution of East Indian sandalwood extends from 30^0 N to 40° S from Indonesia in the east to Juan Fernandez Islands (Chile) in the west and from Hawaiian Archipelago in the north to New Zealand in the south. In India, naturally distributed in the Southern part of India (Karnataka, Tamil Nadu and Kerala) particularly, Karnataka state holds a special place in the culture and heritage since inception so, rightly called as 'the land of sandalwood' (Srinivasan et al. 1992).

The first Sanskrit medical treatises, *Charaka Samhita* and *Sushrata Samhita*, date back to 2000 BC and describe the use of 700 plants, sandalwood one among them (Swerdlow and Johnson 2000). Historical survey reveals that sandalwood has been referred in Indian mythology, folklore and ancient scriptures and cultures place great significance on its fragrance and

M. Gowda et al. (eds.), *The Sandalwood Genome*, Compendium of Plant Genomes,
https://doi.org/10.1007/978-3-030-93394-4_8

medicinal attributes. However its first recorded sandalwood trade commenced much earlier, with the introduction of Buddhism into China from India and it is possible that sandalwood was introduced from Indonesia into India or vice versa. It was a much sought item of trade since the Vedic period, during period of Vijayanagara Empire and Kings of Mysore subsequently (Ganeshaiah et al. 2007). Sandalwood was a priced commodity for both Arabs and Europeans (Nageswara Rao et al. 2001).

The volatile oil yield from a matured tree varies from 3 to 6% depending on various factors like the age of the tree, species, geographical origin and the environment (Shankaranarayana and Kamala 1989) and techniques used for distillation effleurage, steam distillation, solvent extraction, cold pressing and maceration, depending on the stability of the components.

There is significant variation in heartwood chemistry in different parts of the tree. *S. album*, a strong variability in oil content was observed among the different parts of the tree, *i.e.* roots (8.4%), trunk (5.8%) and branches (3.5%). (Srimathi et al. 1995) The heartwood accumulates fragrant sesquiterpenoid santalols until approximately 10–12 years of age and oil content increases significantly with tree age (Jones et al. 2007).

For centuries, sandalwood oil has been a popular ingredient in Ayurvedic medicine as well as traditional Chinese medicine. In recent years, the use of fragrances for clinical purposes has been propagated. The scientific and medicinal benefits of the Indian sandalwood are being rediscovered today in the context of modern medicine (Rajsmita et al. 2019) for its antiviral activity of sandalwood oil against herpes simplex viruses 1 and 2. It was found that the replication of these viruses was inhibited in the presence of the oil (Benencia and Courrèges 1999). The α-santalol has anticancer property by exerting cytotoxic effects against cancer cells through induction of cell cycle arrest and apoptosis (Bommareddy et al. 2019). It has also been found to have anti inflammatory effects by altering the expression of various cytokines and chemokines. However, sandalwood oil has been found to have a myriad of different pharmacological effects and used in the ever-expanding areas of aromatherapy as well. It is one of the valuable inheritances of ancient India to the world of medicine and goes much beyond being a beautifying and fragrance imparting agent. In the field of agriculture, attempts are being made at the identification of eco-friendly natural biocides including essential oils, to replace synthetic pesticides and herbicides. This chapter describes the photochemistry and its importance to devise synthetic molecule and knowledge gap requiring future research.

8.2 Importance of Sandalwood Oil

Sandalwood oil is extensively used in perfume industry due to its coveted long lasting fragrance, sandalwood oil is used as cosmetics, soaps, musk, face packs and food grade oils approved for use as a food flavouring (mainly for chocolates and candies). Sandalwood oil is also frequently incorporated into religious ceremonies or spiritual practices as well. Buddhists believe that the scent of sandalwood can help maintain calmness and focus during meditation. In Ayurvedic medicine, sandalwood oil is believed to be beneficial in treating both physical and mental disorders, including anxiety, bronchitis, diarrhea, fatigue, fever, gallbladder problems, high blood pressure, indigestion, insomnia, liver problems, low libido, sore throat and urinary tract infections.

In aromatherapy, inhaling the aroma of sandalwood oil or absorbing it through the skin is thought to transmit messages to parts of the brain involved in controlling emotions, known as the limbic system. These messages are believed to affect both an emotional and physiological response. Sandalwood oil, when inhaled or applied to the skin, triggers a mild increase in blood oxygen saturation, blood pressure, breathing rate eye blink rate, pulse rate, and skin temperature and α-santalol rendered significant physiological changes which are interpreted in terms of a relaxing or sedative effect (Hongratanaworakit et al. 2004). Farnesol found in

Australian sandalwood (*Santalum spicatum*) enhanced amphotericin B and caspofungin activity against Candida (Cordeiro et al. 2013). It has long antiquity in ethno pharmacology in India and elsewhere, for uses such as a tonic to the immune system. Recent literature reveals sandalwood oil has an antimicrobial property, anticancer properties, cardiotonic, diuretic, diaphoretic, expectorant, aphrodisiac, hemostatic, anodyne and antipyretic properties (Warrier 1993). Sandalwood is one of the potential anti-ulcer agents used in Unani system of medicine. The soothing and demulcent effects of sandalwood oil have been used to treat respiratory tract infections, specifically chronic bronchitis involving chronic dry cough. Recently, apoptotic, cytotoxic, antiviral and anti-*Helicobacter pylori* properties have been reported (Misra and Dey 2013).

8.3 Diversity in Chemical Composition of Sandalwood Oil

Genetic diversity among Sandal populations of different provenances in India as well as relationship between girth and heartwood oil yield was studied (Angadi et al. 1988) (Jain et al. 2003). Although both Indian as well as Australian sandalwood oils contain similar components, the concentrations of these components are different, creating two similar but quite distinctive oils.

Sesquiterpenes (C15) consists of three isoprene units and can be monocyclic, bicyclic, or tricyclic, have a greater potential for stereochemical diversity (Ashour et al. 2010). The structural diversity that arises in the assembly of the 15-carbon skeletons makes up the backbone of all sesquiterpenes and have strong odours. They are found particularly in higher plants (Table 8.1). naturally, they occur as hydrocarbons or in oxygenated forms including lactones, alcohols, acids, aldehydes and ketones. Sesquiterpenes oxidise over time into sesquiterpenols. Main constituents in Australian Sandalwood oil are sesquiterpenes dominated by the two primary sesquiterpene alcohols such as α-santalol and β-santalol along with both E, E-farnesol and α-bisabolol (Table 8.2). Interestingly, dendrolasine (green, fatty and metallic) is up to 200 times more abundant in *Santalum spicatum*, and is probably one of the many minor constituents responsible for the different odour properties between this species and *S. album*.

The quality of sandalwood timber, measured by its oil content, is dependent on its species. *Santalum album* has the highest oil content (~6%, India) followed by *S. yasi* (~5%, Fiji), *S. austrocaledonicum* (~4%, New Caledonia, Australia) and *S. spicatum* (~2%. Australia). The essential oil develops in the heartwood of the stem and root. The geographical location significantly influences the heartwood formation (Demole et al. 1976). There is a compositional difference in oils obtained from young and mature sandal trees and the content and composition of oil varied from heartwood at different levels in the tree (Shankaranarayana and Parthasarathi 1984). Terpene synthase (TPS) genes are expressed in *S. album* which may contribute to the majority of heartwood essential oil (Nageswara Rao et al. 2007). Recent advances in transcriptome and genome sequencing of sandalwood have shed a light on the discovery of many fragrance biosynthetic pathway genes including santalene, bergamotene, curcumene and bisabolene in East Indian Sandalwood oil (Jones et al. 2006).

Sandalwood tree accumulates sesquiterpenes ($C_{15}H_{24}O$), predominantly cis-α-santalol described as slightly woody, reminiscent of cedarwood, α-cedrene and cis- β santalol determines the quality of odour. The α –trans-bergamotol and epi-cis-β-santalol are the major fraction of essential oil (Fig. 8.1). Usually, α-santalol is more abundant than β-santalol (Verghese et al. 1990). Minor components include trans-β-santalol and cis-lanceol, hydrocarbon santene (C_9H_{14}), α-santalene, β-santalene, α-bergamotene, epi-β-santalene, α-curcumene, β-curcumene, γ-curcumene, β-bisabolene, α-bisabolol trace elements found in essential oils which may contribute different olfactory note (Jones et al. 2006). The other constituents reported in the sandalwood oil includes alcohol, santenol ($C_9H_{16}O$) and

Table 8.1 Example of isolated aroma chemicals substance

Substance	Source (Enantiomer)
Santalol	Sandalwood
Limonene	Sweet orange
Citronellol	Geranium
Citral	Lemongrass
Bisabolane	Lavender
Cedrol	Cedarwood

* Adopted from (Ishikawa et al. 2010)

Table 8.2 Major constituents of commercial East Indian sandalwood oils, Australian and New Caledonian oils

Constituent	East Indian Sandalwood Oils	Australian Oils	New Caledonian Oils
(Z)-trans-α-Bergamotol	4.5–8.6	1.0–8.9	5.0–8.6
(Z)-α-Santalol	43.4–53.3	17.0–42.9	38.6–46.6
(E)-β-Santalol	0.5–2.0	0.2–1.3	0.2–0.9
(Z)-β-Santalol	15.6–23.6	6.4–17.7	13.2–19.2

* *Adopted* from (De Groot and Schmidt 2017)

teresantalol ($C_{10}H_{16}O$); the aldehydes, nor-tricycloekasantalal ($C_{11}H_{16}O$) and isovaleraldehyde; the ketones, l-santenone ($C_9H_{14}O$) and santalone ($C_{11}H_{16}O$) and the acids, teresantalic acid ($C_{10}H_{14}O$) occurring partly free and partly in esterified form and α- and β-santalic acids ($C_{15}H_{22}O_2$). *β*-isomer is usually twice less abundant, is more potent and is responsible for the highly prized typical warm-woody, milky, musky, urinous, animal aspects of sandalwood.

More than 230 constituents have been identified so far in the heartwood oil of sandalwood. Interestingly, acetates of santalols and nuciferol in a 80 year-old East Indian essential oil (*S. album*) were not detected in young oils and may be due to *in situ* acetylation of age-related oxidative degradation of sesquiterpene allylic alcohols (Baldovini et al. 2011).

8.4 Conclusion

Santalum species is still an important raw material for chemist and perfumer's interest. In the present review, it is evident that sandalwood oil composition may vary depending on various

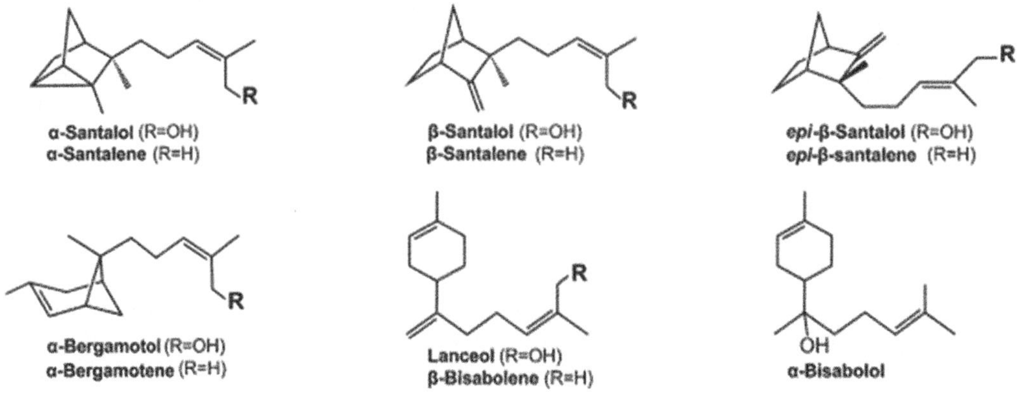

Fig. 8.1 Molecular structure of main sesquiterpenes alcohols/aldehydes of *Santalum album* L

factors like genetic, geographical and taxonomic origin and the method of oil extraction as well. Knowledge on the phytochemistry of the genus *Santalum* is still sceptical and complex. Experimental data is still insufficient to decipher the odorant properties, molecular structures of the chemical components and biosynthetic pathways. In addition, medicinal value in *Santalum* species is getting more value than ever before. However, in recent years, scarce supply of natural heartwood and ever increasing oil price have hampered its use in perfumery industries. In addition, essential oil distillation from particular part is time consuming, energy and labour intensive which restrain the use in essential oil industry. Understanding the phytochemistry and structural complexity of the target molecule will help to develop an industrially feasible synthetic version of the principal odorant molecule and optimisation of the final cost of molecule.

References

Angadi VG, Kamala BS, Rai SN (1988) Effect of deficiency of trace elements on leaf area, chlorophyll level and photosynthetic efficiency in tree-seedlings. Myforest 24:124–128

Ashour M, Wink M, Gershenzon J (2010) Biochemistry of terpenoids: monoterpenes, sesquiterpenes and diterpenes. Annu Plant Rev 40:258–303

Baldovini N, Delasalle C, Joulain D (2011) Phytochemistry of the heartwood from fragrant *Santalum* species: A review. Flavour Fragr J 26:7–26. https://doi.org/10.1002/ffj.2025

Benencia F, Courrèges MC (1999) Antiviral activity of sandalwood oil against Herpes simplex viruses-1 and -2. Phytomedicine 6:119–123. https://doi.org/10.1016/S0944-7113(99)80046-4

Bommareddy A, Brozena S, Steigerwalt J et al (2019) Medicinal properties of alpha-santalol, a naturally occurring constituent of sandalwood oil. Nat Prod Res 33:527–543

Cordeiro RA, Teixeira CEC, Brilhante RSN et al (2013) Minimum inhibitory concentrations of amphotericin B, azoles and caspofungin against *Candida* species are reduced by farnesol. Med Mycol 51:53–59

De Groot AC, Schmidt E (2017) Essential oils, part VI: sandalwood oil, Ylang-Ylang oil, and jasmine absolute. Dermatitis 28:14–21. https://doi.org/10.1097/DER.0000000000000241

Demole E, Demole C, Enggist P (1976) A chemical investigation of the volatile constituents of East Indian sandalwood oil (*Santalum album* L.). Helv Chim Acta 59:737–747

Ganeshaiah KN, Uma R, Vasudeva R (2007) Bioresources and empire building: what favoured the growth of Vijayanagara Empire? Curr Sci 93:140–146

Hongratanaworakit T, Heuberger E, Buchbauer G (2004) Evaluation of the effects of East Indian sandalwood oil and α-Santalol on humans after transdermal absorption. Planta Med 70:3–7. https://doi.org/10.1055/s-2004-815446

Ishikawa M, Warita, Takahisa E, Ohkubo Y (2010) Human–environment interactions (1): flavor and fragrance. In natural products II chemistry and biology. Mander L, Lui HW, Elseviers, Oxford 4:595–629

Jain SH, Angadi VG, Shankaranarayana KH (2003) Edaphic, environmental and genetic factors associated with growth and adaptability of sandal (*Santalum album* L.) in provenances. Sandalwood Res Newsl 17:6–7

Jones CG, Ghisalberti EL, Plummer JA, Barbour EL (2006) Quantitative co-occurrence of sesquiterpenes; a tool for elucidating their biosynthesis in Indian sandalwood, *Santalum album*. Phytochemistry 67:2463–2468

Jones CG, Plummer JA, Barbour EL (2007) Non-destructive sampling of indian sandalwood (*Santalum album* L.) for oil content and composition. J Essent Oil Res 19:157–164. https://doi.org/10.1080/10412905.2007.9699250

Mahesh HB, Subba P, Advani J et al (2018) Multi-omics driven assembly and annotation of the sandalwood (*Santalum album*) genome. Plant Physiol 176:2772–2788. https://doi.org/10.1104/pp.17.01764

Misra BB, Dey S (2013) Quantitative and qualitative evaluation of esquiterpenoids from essential oil and in vitro somatic embryos of east Indian sandalwood (*Santalum album*) tree by HPTLC and GC. Open Access J Med Aromat Plant 4:1–9

Nageswara M, Ganeshaiah KN, Uma R (2007) Assessing threats and mapping sandal resources to identify genetic "hot-spot" for in-situ conservation in peninsular India. Conserv Genet 8:925–935. https://doi.org/10.1007/s10592-006-9247-1

Nageswara M, Ganeshaiah KN, Uma R (2001) Mapping genetic diversity of sandal (*Santalum album* L.) in south India: lessons for *in-situ* conservation of sandal genetic resources. In: Uma R, Ganeshaiah KN, Bawa KS (eds) Forest genetic resources: status, threats and conservation strategies. Oxford and IBH Publications, New Delhi, India, pp 49–67

Rajsmita B, Keshavamurthy V (2019) Re-discovering sandalwood: Beyond beauty and fragrance. Indian Dermatol Online J 10(3):296

Shankaranarayana KH, Kamala BS (1989) Fragrant products from less odorous sandalwood oil. Perfum Flavorist 14:19–20

Shankaranarayana KH, Parthasarathi K (1984) Synthetic sandalwood aroma chemicals perfum. Flavor 9:17–20

Srimathi RA, Kulkarni HD, Venkatesan K (1995) Recent advances in research and management of sandal

(*Santalum album* L.) in India. Associated publishing Co, New Delhi
Srinivasan VV, Sivaramakrishnan VR, Rangaswamy CR, Ananthapadmanabha HS, Shankaranarayana KH (1992) Sandal (*Santalum album* L.). Indian Council of Forestry Research and Education, Dehradun, India
Subasinghe SMCUP (2013) Sandalwood research: a global perspective. J Trop For Environ 3. https://doi.org/10.31357/jtfe.v3i1.1117
Swerdlow JL, Johnson L (2000) Nature's medicine plants that heal. National Geographic Society, Washington, D.C.
Verghese J, Sunny TP, Balakrishnan KV (1990) Alpha santalol and beta santalol, a new quality determination of East Indian sandalwood oil. Flavour Fragrance 5:223–226
Warrier PK, Nambiar VPK, Ramankutty C (1993) Indian medicinal plants: a compendium of 500 species. Orient Longman Publishers, Kottakkal, India 1–5

9 Scope of Proteogenomic Analysis for Indian Sandalwood (*Santalum Album*)

Pratigya Subba and Thottethodi Subrahmanya Keshava Prasad

Abstract

Sandalwood oil is recognized as the queen of essential oils due to its highly valued fragrance. Globally, 16 species of sandalwood have been described among which the East Indian sandalwood, *Santalum album* is appreciated for superior quality and quantity of essential oils. The oils are extracted from the central dead heartwood and roots of this hemiparasitic tropical tree. Utility of sandalwood oil has been described in Ayurveda as an antiseptic and astringent. India remains the global leader in sandalwood oil production supplying 95% of the world's requirement. Originally, harvests were made from the wild, which eventually led to over-exploitation of forests. Due to its diminishing population, *S. album* has been declared as a 'vulnerable' species. Appropriate management strategies are required for its conservation and germplasm preservation. Owing to its immense economical importance, it is imperative to generate an in depth molecular map of *S. album*. Integrative multi-omics data enable comprehensive understanding of biological networks and their interrelation. Comprehensive proteomic analysis provides insights into protein expression levels, allows identification of several types of post-translational modifications on proteins and aids in genome annotation. Proteogenomics workflow efficiently combines datasets derived from proteomic, genomic, and transcriptomic analyses to discover novel protein-coding genes and improve the computationally derived genome annotation. Here, we discuss the status and scope of proteogenomics in continued efforts of genome annotation of *S. album*.

P. Subba · T. S. K. Prasad (✉)
Center for Systems Biology and Molecular Medicine, Yenepoya Research Center, Yenepoya (Deemed to be University), Mangalore 575018, India
e-mail: keshav@yenepoya.edu.in

P. Subba
e-mail: pratigyas@yenepoya.edu.in

9.1 Introduction

Santalum album popularly known as the Indian Sandalwood is one of the most expensive woods valued for its fragrant essential oils. This tropical semi-parasitic tree is native to India and Australia, but also grows in China, Indonesia, and the Philippines. The growth conditions include hot and humid conditions and require very little water for growth. The plant belongs to family *Santalaceae* and genus *Santalum*. The tree grows to a height of 33–66 feet and the colour of the wood varies from dark brown, reddish to yellow–brown. The plant is harvested for its wood that serves as an excellent carving material and the heartwood and roots that produce high in demand sandalwood oil. In addition to its

M. Gowda et al. (eds.), *The Sandalwood Genome*, Compendium of Plant Genomes,
https://doi.org/10.1007/978-3-030-93394-4_9

commercial value, in Ayurveda system of medicine, it is used predominantly to treat skin ailments and as an antiseptic due to its soothing and natural astringent properties. Sandalwood in India also holds high reverence due to its religious sanctity in the Hindu and Buddhist religions.

In India, growth of sandalwood once spread across the southern peninsula but is now restricted majorly to two states, Karnataka and Tamil Nadu. Referred to as the 'pride of Karnataka's forest wealth', Karnataka produces approximately 80% of sandalwood oil in India. *S. album* is also assigned as the state tree of Karnataka. The 'vulnerable' status declared by the International Union for the Conservation of Nature's (IUCN) Red List demands strategizing methods for the conservation of *S. album*. Molecular data provides crucial knowledge about genetic diversity and identification of markers that can be leveraged for designing conservation methods. Draft genome sequences and transcriptome sequences have been described for Sandalwood using Next-generation sequencing platforms. However, like many other genome sequences, genome annotation remained far from complete. Proteogenomic protocols, an integrative approach of using proteomics data for the reannotation of the genome have allowed validation of a large number of predicted gene models at protein level, and identification of novel protein-coding genes missed by computational approaches generally used in genome annotation (Renuse et al. 2011). Inclusion of this workflow has improved knowledge of the genome structure in several organisms including humans (Nesvizhskii 2014; Renuse et al. 2011). Here, we discuss the merger of genomics, transcriptomics and proteomics data for robust annotation of the *S. album* genome.

9.2 Proteogenomics: A Leeway to Refine Genome Annotation

High resolution mass spectrometry allows identification of several thousand proteins in a single experiment. In a typical proteomics data analysis set up, MS/MS data is searched against the protein database provided by the genome sequencing team. In other words, those who analyse proteome data assume that the protein sequence database contains all proteins/peptides of the organism being investigated (Nesvizhskii 2014). However, new protein forms from alternate splice forms, mutated genes or novel genes, which are missed by computational strategies, are not part of the protein sequence databases. As a result, even though data for these proteins are captured during MS/MS analysis, identification of these novel events cannot be extracted using typical proteomic analysis strategies. In order to overcome these loopholes, new proteomic analysis strategies called 'Proteogenomics' was developed. Here, the mass spectrometric data that remain unassigned when searched against the protein database are extracted and searched against a custom protein databases generated by the six-frame translation of genome and/or three-frame translation of transcript database and long noncoding RNAs data among others (Kim et al. 2014). This enables the identification of novel genes and aids refinement of the annotated gene structures.

Previously, proteogenomics pipelines have been used to improve the genome annotation of several organisms (Table 9.1). In the human genome, proteogenomics enhanced the representation of protein-coding genes together with identification of coding ability of long intergenic noncoding RNAs (Kim et al. 2014; Wilhelm et al. 2014). Proteogenomics, when applied to proteomic data generated for cancer-derived samples, has allowed identification of potential novel biomarkers and therapeutic targets (Hernandez-Valladares et al. 2017). In the vector mosquito *Anopheles stephensi*, proteogenomics allowed incorporation of substantial corrections in the genome including the identification of 365 novel genes and conversion of 87 predicted 'noncoding RNAs' to m-RNAs of protein-coding genes (Prasad et al. 2017). These revised genome annotations were further extended to 15 other Anopheles species. Despite being the most well worked upon *Mycobacterium tuberculosis* strain, deep proteomics analysis of H37Rv allowed delineation of 41 novel genes and corrected 79 gene models including revisions of translation start sites

(Kelkar et al. 2011). In the attenuated *M. tuberculosis* strain H37Ra, multi-OMICs analysis led to the identification of 65 novel protein-coding genes and correction of 25 existing gene models (Pinto et al. 2018). Among plants, in *Zea mays,* several hundred modifications in gene annotation were proposed through proteogenomics (Castellana et al. 2014).

9.3 *S. Album*: Foray of into—Omics Era

Next-generation sequencing methods are high-throughput techniques that provide nucleotide information of thousands of genes along with identification of mutations, polymorphisms, structural variants, etc. Elucidation of the genetic components involved various biosynthetic pathways especially the biosynthesis of essential oils are of utmost consideration in sandalwood. In *S. album*, genomic data will enable the identification of genes underlying phenotypic differences and differences associated with essential oil production compared to Australian sandalwood *S. spicatum*, another sandalwood variety traded internationally. Importantly, *S. spicatum* is also known to produce lower yield of the essential oils. Therefore, genomic data will also allow researchers to ascertain genes associated with yield parameters. *S. album* is a hemiparasite that depends on a variety of host plants for its nitrogen supplementation (Lu et al. 2014). External factors some of which may include host derived molecules are also reported to drive growth in *S. album* (Li et al. 2021). However, exact molecular mechanisms that regulate hemiparasitism in *S. album* are not fully known. Data generated from genomic studies will provide information regarding these processes that may help develop strategies to overcome host dependency in *S. album*.

Our team reported the draft genome of *S. album* together with transcriptome and proteome data. Flow cytometric investigations of *S. album* showed the diploid somatic chromosome number is $2n = 20$ (Mahesh et al. 2018). The estimated genome size is 221 Mb, one of the smallest genome size reported for a woody plant (Mahesh et al. 2018). We reported that the genome encodes 38,119 protein-coding genes and is composed of 27.42% repetitive DNA elements. Following our report, another article on genome sequencing of *S. album* was published (Dasgupta et al. 2019). Terpenoids are reported as major constituents of aromatic compounds associated with flavour. Terpene synthases (TPSs) are of prime importance in terpenoid biosynthetic pathways and few TPSs have been characterized in sandalwood with respect to their roles in sandalwood oil biosynthetic pathway (Zhang et al. 2019). In *S. album*, according to one study 30 genes encoding TPSs were identified (Dasgupta et al. 2019). In addition to *S. spicatum*, essential oils are also extracted from other sandalwood species including *S. yasi*, and *S. asutrocaledonicum* (Thomson 2020). In future, large-scale comparative genome analysis will also enable estimation of genetic diversity among various other sandalwood species. In addition to the genome sequence, the chloroplast genome 'the plastome' of *S. album* has also been sequenced (Guo et al. 2020). Reportedly, the chloroplast genome is 144,101 bp in length and encodes for 123 genes, including 80 protein-coding genes, 8 ribosomal RNA genes, and 35 transfer RNA genes (Yang et al. 2020). Our team also reported the identification of 25 mitochondrial genes (Mahesh et al. 2018). In addition to the *S. album* genomics data, we generated the transcriptome and high-resolution mass spectrometry based proteomics data to: (i) validate the predicted gene models, (ii) conduct tissue-specific expression analysis, and (iii) improve genome annotation.

9.4 Validating the Predicted Gene Models of *S. Album*

Utilizing a multi-omics strategy, our group reported the genome sequence of *S. album* that was further supported by the transcriptome and high-resolution mass spectrometry-derived proteome data (Mahesh et al. 2018). Transcriptome represents the complete range of coding and noncoding RNA molecules in a cell type, tissue or developmental stage of an organism at a certain time. In our data, of the 38,061 predicted genes, 29,212 genes were associated with evidence from RNA-Seq data. This parallel analysis of the *S. album* genome and transcriptome enabled the validation of coding potential of the predicted gene models at the transcriptome level. In addition to our analysis of the *S. album* transcriptome, another interesting study reported differential expression of genes related to sandalwood oil biosynthesis in 3 zones of the mature wood. These oils are extracted from the central dead wood known as the heartwood as the contents are higher than in the outer sapwood (Brand et al. 2012). Although the sandalwood heartwood remains the main source of these oils, roots are also a rich source of the oils. The percentage and weight of heartwood varies based on the age and importantly the oil content also varies in the heartwood of juvenile and aged trees (Arun Kumar et al. 2011). Based on the hypothesis that the expression patterns of santalene/bergamotene hydroxylases, the key enzymes involved the biosynthesis of sesquiterpenols vary in the sapwood and the heartwood an RNA-Seq analysis was carried out (Celedon et al. 2016). For the analysis, transcriptomes of the sapwood (SW), transition zone (TZ) and heartwood (HW) were compared. The 3 types of wood shavings were collected based on the colour demarcations that ranged from pale yellow to pink and dark brown, respectively. The analysis revealed differential expression of transcripts among the 3 wood types. Unique sesquiterpenoid transcriptome signatures were highly expressed in the HW including mevalonic acid (MEV) pathway, and farnesyl diphosphate synthase (FPPS) transcripts. Notably, sesquiterpenols make up to 90% of the sandalwood essential oils (Diaz-Chavez et al. 2013). A total of 116 unique transcripts encoding predicted P450 proteins were also highly expressed in HW. The function of one of these genes *SaCYP736A167* was also tested for its ability to convert α-santalene into (Z)-α-santalol and β-santalene into (Z)-β-santalol. Transcriptome analysis of the *S. spicatum* heartwood has also been reported wherein a number of sesquiterpenoid biosynthetic pathway associated genes (Moniodis et al. 2015). More recently, comparative transcriptome analysis of genetically identical sandalwood trees assigned expression of cytochrome family of genes to high and low oil yielding trees (Fatima et al. 2021). In order to use nutrients from host plants, sandalwood uses haustoria to access the host's xylem sap (Zhang et al. 2015). One study reported the transcriptome analysis of genes associated with haustorial development of *S. album* and identified genes encoding nodulin-like proteins that may be key haustorial morphogenesis (Zhang et al. 2015). We also reported the transcript evidence for 10 genes encoded by the mitochondrial genome (Mahesh et al. 2018). The diverse studies were conducted to explore various aspects of *S. album* biology.

In order to perform in depth proteomics analysis, we extracted and analyzed proteins from 4 tissues of *S. album* viz., shoot meristem, leaf, fruit and stems (Mahesh et al. 2018). The proteins were digested overnight using the proteolytic enzyme trypsin. Tryptic peptides derived from each tissue were fractionated into 96 fractions using basic reverse-phase liquid chromatography (bRPLC). The fractions were then concatenated into 10 fractions. Each fraction was analyzed separately using high-resolution Orbitrap Fusion Tribrid mass spectrometer coupled to an EASY-nLC 1200 system. The analysis led to the identification of 72,325 peptides that mapped to 10,076 genes predicted in the sandalwood genome. For this part of analysis, the protein database was generated from the genome assembly, MAKER-P-predicted genes with RNA-seq evidence and MAKER-P-predicted ab initio method. The identifications of these peptides/proteins validated the expression of the predicted gene models. Protein quantification across the 4 tissues was calculated using an intensity-based absolute quantification (iBAQ) method. Tissue specific expression patterns were observed at the protein level with ribulose bisphosphate carboxylase (small chain) and chlorophyll a/b-binding protein identified as the most abundant proteins in the leaf tissue. In the stem, abundantly expressed proteins included cyclophilin-type peptidyl-prolyl cis-trans-isomerase/CLD, uncharacterized protein containing Pfam domain ABA/WDS-induced protein, uncharacterized protein containing Pfam domain pathogenesis-related protein Bet v I family and uncharacterized protein containing Pfam domain aldo/keto-reductase family. In the multi-OMICs analysis of *S. album*, our group reported the identification of 184 cytochrome P450 genes in the genome the evidence of which were also supported by RNA-Seq and proteogenomics data (Mahesh et al. 2018). These findings pave way for the complete understanding of sandalwood terpenoid biosynthesis. From the perspective of the plant, essential oils are primarily synthesized to prevent decaying of the wood. The role of CYP76 family of cytochromes P450 proteins in santalol biosynthesis was characterized by individually cloning 10 major isoforms and assessment of their sesquiterpene oxidase activity (Diaz-Chavez et al. 2013; Jones et al. 2011). However, activity of these enzymes produced very low percentage of the (Z) stereoisomers and majorly led to the production of (E)-α-santalol, (E)-β-santalol, (E)-epi-β-santalol and (E)-α-exo-bergamotol. An opportunity for identification of more cytochromes P450 enzymes and exploring their involvement in the biosynthesis of essential oils remains open. Subsequently, Celedon et al. have reported the characterization of *SaCYP736A167* that was distinct from previously described cytochromes P450 proteins and produced (Z)-α-santalol, (Z)-β-santalol, (Z)-epi-β-santalol and (Z)-α-exo-bergamotol in yeast cells (Celedon et al. 2016).

9.5 A Case Study: Proteomics Assisted Genome Annotation of *S. Album*

Using the massively high-throughput Next Generation Sequencing technique, data for several thousand organisms have been generated so far. Popularity of this technique lies in its high accuracy, reproducibility and a priori knowledge about the genomic features and increased affordability. Genomic data has now been incorporated in clinical settings for detection of clinically actionable mutations with potential to design 'personalized medicine'. Annotation of the genome and transcriptome data are performed computationally using gene prediction tools that are programmed to perform either similarity based searches using ab initio methods. A stumbling block in genome annotation is the errors in protein-coding gene prediction. According to one study, the errors affected approximately 50% of the sequence (Meyer et al. 2020) Reannotation of the genomes using different software tools leave ample space for identification of novel genes and fix errors indicating imprecise strategies of genome annotation.

In our study, we searched the unassigned MS/MS data against a custom six-frame translated sandalwood genome database and three-frame translated sandalwood transcript database to identify novel events (Fig. 9.1). Peptides that mapped exclusively to these custom databases with no prior evidence for their protein-coding capability were termed as genome search-specific peptides (GSSPs). Further, a set of high-confident GSSPs were obtained by applying the following filters: (1) Xcorr (cross correlation) value above 3 and (2) ion score above 30. A set of high-confident 53 novel protein-coding genes were identified, 46 of which were also supported by transcripts from a RNA-seq analysis. Interestingly, a significant number of the novel genes consisted of uncharacterized proteins that serve as interesting candidates for further functional characterization. We also reported the identification of long noncoding RNA (lncRNA) transcripts 1,348 of which were also supported by peptide evidence from our mass spectrometry-derived proteomics data. The proteogenomics analysis also allowed the revision of several annotated gene structures of sandalwood.

Investigation of genome structure and composition using integrated omics approach clearly provides an immeasurable amount of information that cannot otherwise be provided by genome sequencing method alone. In this chapter, we presented enhanced genome annotation of a nonmodel plant *S. album* using a combinatorial strategy including genomics, transcriptomics, proteomics and proteogenomics. The genome

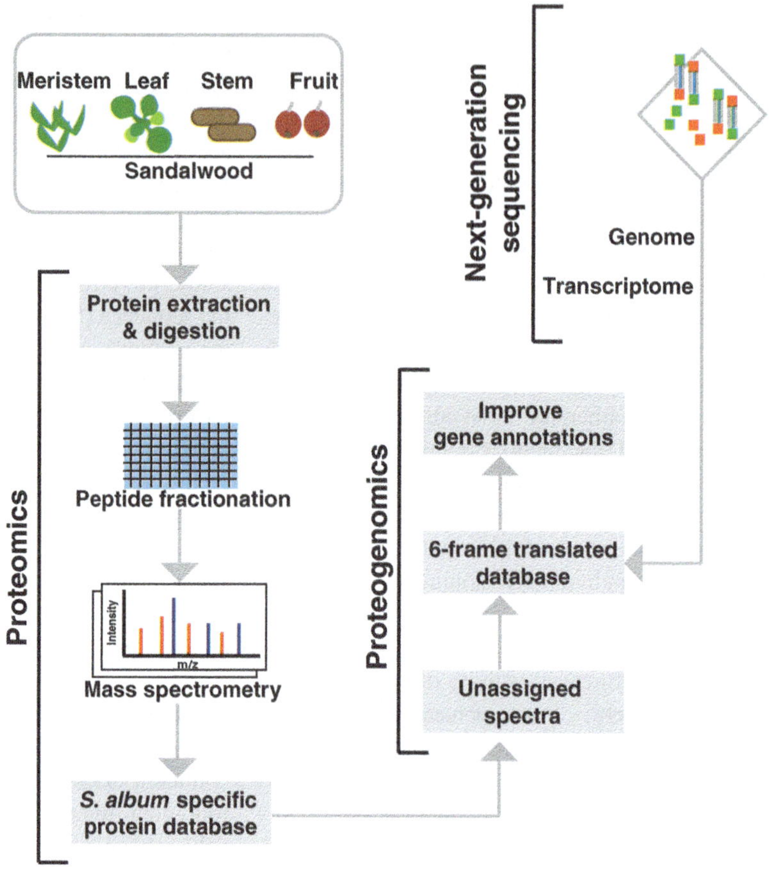

Fig. 9.1 Schematic workflow for proteogenomic analysis of sandalwood tissues. The genome and transcriptome data were generated using Next-generation sequencing. Mass spectrometry-derived proteomics data were searched against the protein database, six-frame translated genome database and three-frame translated transcriptome database

annotation is well supported by the transcriptomics and proteomics data. The seamless merger of three datasets and simultaneous ability to view them in a single window serves as an auxiliary mechanism to refine genome annotations. The draft genome of *S. album* presents enormous opportunities to decode the biosynthetic pathways operative in this economically important plant. Additionally, future endeavours in comparative molecular analysis will delineate species specific differences in various sandalwood species. The introduction of similar workflows for genome annotation will assist computational annotation programmes resulting in unravelling of robust genome structures with potentiality for making appropriate biological interpretations.

Table 9.1 List of proteogenomics studies carried out in various organisms

Article	Organism	Authors	Journal
Genome annotation of *Anopheles gambiae* using mass spectrometry-derived data	*Anopheles gambiae*	(Kalume et al. 2005)	*BMC Genomics*
Genome-scale proteomics reveals *Arabidopsis thaliana* gene models and proteome dynamics	*Arabidopsis thaliana*	(Baerenfaller et al. 2008)	*Science*
Discovery and revision of Arabidopsis genes by proteogenomics	*Arabidopsis thaliana*	(Castellana et al. 2008)	*Proc Natl Acad Sci U S A*
A high-quality catalog of the *Drosophila melanogaster* proteome	*Drosophila melanogaster*	(Brunner et al. 2007)	*Nat Biotechnol*
Use of shotgun proteomics for the identification, confirmation and correction of *C. elegans* gene annotations	*Caenorhabditis elegans*	(Merrihew et al. 2008)	*Genome Res*
Proteogenomic analysis of *Mycobacterium tuberculosis* by high-resolution mass spectrometry	*Mycobacterium tuberculosis*	(Kelkar et al. 2011)	*Mol Cell Proteomics*
Proteogenomic analysis of *Candida glabrata* using High Resolution Mass Spectrometry	*Candida glabrata*	(Prasad et al. 2012)	*J Proteome Res*
Reconstruction of protein networks from an atlas of maize seed proteotypes	*Zea mays*	(Walley et al. 2013)	*Proc Natl Acad Sci U S A*
An automated proteogenomic method uses mass spectrometry to reveal novel genes in *Zea mays*	*Zea mays*	(Castellana et al. 2014)	*Mol Cell Proteomics*
A draft map of the human proteome	*Homo sapiens*	(Kim et al. 2014)	*Nature*
Mass-spectrometry-based draft of the human proteome	*Homo sapiens*	(Wilhelm et al. 2014)	*Nature*
A high-resolution tissue-specific proteome and phosphoproteome atlas of maize primary roots reveals functional gradients along the root axes	*Zea mays*	(Marcon et al. 2015)	*Plant Physiol*
A proteomic atlas of the legume *Medicago truncatula* and its nitrogen-fixing endosymbiont *Sinorhizobiummeliloti*	*Medicago truncatula*	(Marx et al. 2016)	*Nat Biotechnol*
Annotation of the zebrafish genome through an integrated transcriptomic and proteomic analysis	*Danio rerio*	(Kelkar et al. 2014)	*Mol Cell Proteomics*
Neglected tropical diseases and omics science: proteogenomics analysis of the promastigote stage of Leishmania major parasite	*Leishmania major*	(Pawar et al. 2014)	*OMICS: A Journal of Integrative Biology*
Proteogenomics of *Candida tropicalis*—an opportunistic pathogen with importance for global health	*Candida tropicalis*	(Datta et al. 2016)	*OMICS: A Journal of Integrative Biology*
Integrating transcriptomic and proteomic data for accurate assembly and annotation of genomes	*Anopheles stephensi*	(Prasad et al. 2017)	*Genome Res*
Proteogenomic analysis greatly expands the identification of proteins related to reproduction in the apogamous fern *Dryopteris affinis*ssp. Affinis	*Dryopteris affinis*	(Grossmann et al. 2017)	*Front Plant Sci*
Integrated multi-omic analysis of *Mycobacterium tuberculosis* H37Ra redefines virulence attributes	*Mycobacterium tuberculosis*	(Pinto et al. 2018)	*Front Microbiol*
Full-length transcript-based proteogenomics of rice improves its genome and proteome annotation	*Oryza sativa*	(Chen et al. 2020)	*Plant Physiol*
Mass-spectrometry-based draft of the *Arabidopsis* proteome	*Arabidopsis thaliana*	(Mergner et al. 2020)	*Nature*

References

Arun Kumar AN, Srinivasa YB, Joshi G, Seetharam A (2011) Variability in and relation between tree growth, heartwood and oil content in sandalwood (*Santalum album* L.). Curr Sci 100:827–830

Baerenfaller K, Grossmann J, Grobei MA, Hull R, Hirsch-Hoffmann M et al (2008) Genome-scale proteomics reveals *Arabidopsis thaliana* gene models and proteome dynamics. Science 320:938–941

Brand JE, Norris LJ, Dumbrell IC (2012) Estimated heartwood weights and oil concentrations within 16-year-old Indian sandalwood (*Santalum album*) trees planted near Kununurra, Western Australia. Aust for 75:225–232

Brunner E, Ahrens CH, Mohanty S, Baetschmann H, Loevenich S et al (2007) A high-quality catalog of the *Drosophila melanogaster* proteome. Nat Biotechnol 25:576–583

Castellana NE, Payne SH, Shen Z, Stanke M, Bafna V, Briggs SP (2008) Discovery and revision of *Arabidopsis* genes by proteogenomics. Proc Natl Acad Sci USA 105:21034–21038

Castellana NE, Shen Z, He Y, Walley JW, Cassidy CJ et al (2014) An automated proteogenomic method uses mass spectrometry to reveal novel genes in *Zea mays*. Mol Cell Proteom 13:157–167

Celedon JM, Chiang A, Yuen MM, Diaz-Chavez ML, Madilao LL et al (2016) Heartwood-specific transcriptome and metabolite signatures of tropical sandalwood (*Santalum album*) reveal the final step of (Z)-santalol fragrance biosynthesis. Plant J 86:289–299

Chen MX, Zhu FY, Gao B, Ma KL, Zhang Y et al (2020) Full-length transcript-based proteogenomics of rice improves its genome and proteome annotation. Plant Physiol 182:1510–1526

Dasgupta MG, Ulaganathan K, Dev SA, Balakrishnan S (2019) Draft genome of *Santalum album* L. provides genomic resources for accelerated trait improvement. Tree Genet Genom 15:34

Datta KK, Patil AH, Patel K, Dey G, Madugundu AK et al (2016) Proteogenomics of *Candida tropicalis* -an opportunistic pathogen with importance for global health. OMICS 20:239–247

Diaz-Chavez ML, Moniodis J, Madilao LL, Jancsik S, Keeling CI, et al. (2013) Biosynthesis of sandalwood oil: *Santalum album* CYP76F cytochromes P450 produce santalols and bergamotol. PLoS One 8: e75053

Fatima T, Krishnan R, Srivastava A, Hanur VS, Rao MS (2021) Comparative transcriptome profiling of high and low oil yielding *Santalum album* L. bioRxiv

Grossmann J, Fernandez H, Chaubey PM, Valdes AE, Gagliardini V, et al. (2017) Proteogenomic analysis greatly expands the identification of proteins related to reproduction in the apogamous fern *Dryopteris affinis* ssp. affinis. Front Plant Sci 8:336

Guo X, Liu C, Zhang G, Su W, Landis JB, et al. (2020) The complete plastomes of five hemiparasitic plants (*Osyris wightiana*, *Pyrularia edulis*, *Santalum album*, viscum liquidambaricolum, and V. ovalifolium): comparative and evolutionary analyses within *Santalales*. Front Genet 11:597

Hernandez-Valladares M, Vaudel M, Selheim F, Berven F, Bruserud O (2017) Proteogenomics approaches for studying cancer biology and their potential in the identification of acute myeloid leukemia biomarkers. Expert Rev Proteom 14:649–663

Jones CG, Moniodis J, Zulak KG, Scaffidi A, Plummer JA et al (2011) Sandalwood fragrance biosynthesis involves sesquiterpene synthases of both the terpene synthase (TPS)-a and TPS-b subfamilies, including santalene synthases. J Biol Chem 286:17445–17454

Kalume DE, Peri S, Reddy R, Zhong J, Okulate M et al (2005) Genome annotation of anopheles gambiae using mass spectrometry-derived data. BMC Genomics 6:128

Kelkar DS, Provost E, Chaerkady R, Muthusamy B, Manda SS et al (2014) Annotation of the zebrafish genome through an integrated transcriptomic and proteomic analysis. Mol Cell Proteom 13:3184–3198

Kelkar DS, Kumar D, Kumar P, Balakrishnan L, Muthusamy B, et al. (2011) Proteogenomic analysis of *Mycobacterium tuberculosis* by high resolution mass spectrometry. Mol Cell Proteom 10:M111 011627

Kim MS, Pinto SM, Getnet D, Nirujogi RS, Manda SS, Annotation of the zebrafish genome through an integrated transcriptomic and proteomic analysis (2014) A draft map of the human proteome. Nature 509:575–581

Li Y, Zhang X, Cheng Q, Teixeira da Silva JA (2021) Annotation of the zebrafish genome through an integrated transcriptomic and proteomic analysis. Elicitors modulate young sandalwood (*Santalum album* L.) growth, heartwood formation, and concrete oil synthesis. Plants (Basel) 10:339

Lu JK, Xu DP, Kang LH, He XH (2014) Host-species-dependent physiological characteristics of hemiparasite *Santalum album* in association with N2-fixing and nonN2-fixing hosts native to southern China. Tree Physiol 34:1006–1017

Mahesh HB, Subba P, Advani J, Shirke MD, Loganathan RM et al (2018) Multi-omics driven assembly and annotation of the sandalwood (*Santalum album*) genome. Plant Physiol 176:2772–2788

Marcon C, Malik WA, Walley JW, Shen Z, Paschold A et al (2015) A high-resolution tissue-specific proteome and phosphoproteome atlas of maize primary roots reveals functional gradients along the root axes. Plant Physiol 168:233–246

Marx H, Minogue CE, Jayaraman D, Richards AL, Kwiecien NW et al (2016) A proteomic atlas of the legume *Medicago truncatula* and its nitrogen-fixing endosymbiont *Sinorhizobium meliloti*. Nat Biotechnol 34:1198–1205

Mergner J, Frejno M, List M, Papacek M, Chen X et al (2020) Mass-spectrometry-based draft of the *Arabidopsis* proteome. Nature 579:409–414

Merrihew GE, Davis C, Ewing B, Williams G, Kall L et al (2008) Use of shotgun proteomics for the identification, confirmation, and correction of *C. elegans* gene annotations. Genome Res 18:1660–1669

Meyer C, Scalzitti N, Jeannin-Girardon A, Collet P, Poch O, Thompson JD (2020) Understanding the causes of errors in eukaryotic protein-coding gene prediction: a case study of primate proteomes. BMC Bioinformatics 21

Moniodis J, Jones CG, Barbour EL, Plummer JA, Ghisalberti EL, Bohlmann J (2015) The transcriptome of sesquiterpenoid biosynthesis in heartwood xylem of Western Australian sandalwood (*Santalum spicatum*). Phytochemistry 113:79–86

Nesvizhskii AI (2014) Proteogenomics: concepts, applications and computational strategies. Nat Methods 11:1114–1125

Pawar H, Renuse S, Khobragade SN, Chavan S, Sathe G et al (2014) Neglected tropical diseases and omics science: proteogenomics analysis of the promastigote stage of *Leishmania* major parasite. OMICS 18:499–512

Pinto SM, Verma R, Advani J, Chatterjee O, Patil AH et al (2018) Integrated multi-omic analysis of *Mycobacterium tuberculosis* H37Ra redefines virulence attributes. Front Microbiol 9:1314

Prasad TSK, Harsha HC, Keerthikumar S, Sekhar NR, Selvan LD et al (2012) Proteogenomic analysis of *Candida glabrata* using high resolution mass spectrometry. J Proteome Res 11:247–260

Prasad TSK, Mohanty AK, Kumar M, Sreenivasamurthy SK, Dey G et al (2017) Integrating transcriptomic and proteomic data for accurate assembly and annotation of genomes. Genome Res 27:133–144

Renuse S, Chaerkady R, Pandey A (2011) Proteogenomics Proteomics 11:620–630

Thomson LAJ (2020) Looking ahead – global sandalwood production and markets in 2040, and implications for Pacific Island producers. Aust for 83:245–254

Walley JW, Shen Z, Sartor R, Wu KJ, Osborn J et al (2013) Reconstruction of protein networks from an atlas of maize seed proteotypes. Proc Natl Acad Sci USA 110:E4808-4817

Wilhelm M, Schlegl J, Hahne H, Gholami AM, Lieberenz M et al (2014) Mass-spectrometry-based draft of the human proteome. Nature 509:582–587

Yang D, Qiu Q, Xu L, Xu Y, Wang Y (2020) The complete chloroplast genome sequence of *Santalum album*. Mitochondrial DNA B Resour 5:406–407

Zhang X, Berkowitz O, Teixeira da Silva JA, Zhang M, Ma G, Whelan J, Duan J (2015) RNA-Seq analysis identifies key genes associated with haustorial development in the root hemiparasite *Santalum album*. Front Plant Sci 6:661

Zhang X, Niu M, Teixeira da Silva JA, Zhang Y, Yuan Y et al (2019) Identification and functional characterization of three new terpene synthase genes involved in chemical defense and abiotic stresses in *Santalum album*. BMC Plant Biol 19:115

Future Prospects of Sandalwood Research

10

H. B. Mahesh, K. G. Manasa, M. K. Prasannakumar, Meghana Deepak Shirke, Chittaranjan Kole, and Malali Gowda

Abstract

Sandalwood is a hemi-parasitic commercially and culturally important tree, mostly valued for the oil and fragrant heartwood. Sandalwood is a forest produce and the second most expensive wood in the world, whereas recent advancements of Next generation sequencing (NGS) technologies considerably promoted forest tree genomics. Genotyping interventions are most important in capturing the natural variability, however draft genome assembly facilitated the sandalwood genomics in developing sequence based markers through GBS, DArT, ddRAD and WGRS which can be utilized in genome-wide association studies, marker assisted selection and genomic selection to develop genetically improved saplings for commercial cultivation.

H. B. Mahesh (✉) · K. G. Manasa
Department of Genetics and Plant Breeding, College of Agriculture, Mandya, University of Agricultural Sciences, Bangalore 560065, India

M. K. Prasannakumar
Department of Plant Pathology, College of Agriculture, GKVK, University of Agricultural Sciences, Bangalore 560065, India

M. D. Shirke
Department of Biotechnology, KTHM College, Nashik 422002, India

C. Kole
ICAR-National Research Center On Plant Biotechnology, Pusa, New Delhi 110012, India

M. Gowda
The University of Trans-Disciplinary Health Sciences and Technology (TDU), Bangalore 560064, India

M. Gowda
Bengaluru Genomics Center, Yelahanka, Bangalore 560064, India

10.1 Introduction

Indian sandalwood (*Santalum album* L.), is considered to be the queen among sixteen recognized sandalwood species of the genus *Santalum* belonging to the family Santalaceae. India stood has the single largest supplier of sandalwood oil meeting more than 80% of the world's demand over past decades. Overharvesting, illegal trading and poaching from its natural habitat resulted in declined area of cultivation and made sandalwood to get classified as endangered plant species in India.

Indian government included the sandalwood as one among other 32 medicinal plants whereas the legal hassles of sandalwood trade regulation still persist. As the supply of Indian sandalwood is inadequate and expensive, industries are looking for alternatives like Australian Sandalwood oil where Santalol content is 38–39% when compared to Indian sandalwood oil which is 90%. Indian Sandalwood oil plays vital role in perfumery industries with its alpha-santalol which is 50–60% and beta santalol which is 15

M. Gowda et al. (eds.), *The Sandalwood Genome*, Compendium of Plant Genomes,
https://doi.org/10.1007/978-3-030-93394-4_10

to 25% as a fixative, modifier and enduring fragrance ingredient. Due to its high-quality oil content and demand at the global market countries like Australia, Sri Lanka, China, Thailand, Cambodia, Costa Rica, etc., are undertaking the huge cultivation projects of Indian sandalwood (http://www.fao.org/3/ap001e/ap001e15.pdf).

10.2 Future Prospects

10.2.1 Population Diversity Studies

Supply of genetically improved quality saplings for a large scale cultivation of sandalwood is a bottle neck. Advancement of genomics and availability of draft genome assembly of sandalwood facilitate to undertake many advanced studies to understand genomics and biology of sandalwood (Mahesh et al. 2018). Sandalwood is a forest produce and application of genotyping technologies in diversity studies for a tree species is important for utilizing the existing natural variability. With the intervention of advanced next generation sequencing (NGS) technologies, utilization of many sequencing platforms have been initiated to perform genotyping. To comprehend the gap of genetic diversity, high-throughput genotyping platforms/technology like genotyping by sequencing (GBS), double digested restriction-site associated DNA (ddRAD) sequencing, diversity array technology (DArT) and whole genome re-sequencing (WGRS) can be used in sandalwood.

10.2.2 Disease Resistance Breeding

Sustainable cultivation techniques are essential for large-scale production and propagation of disease-free *S. album* trees. Many diseases and pests affect *Santalum album*, the most severe of which being '*Candidatus* phytoplasma' causing sandal spike disease (SSD). SSD has been studied for over 120 years and it was first documented in Coorg, South India, in 1899. Phytoplasma is the smallest prokaryotic organism capable of independent reproduction and it is a pleomorphic microorganism (i.e. does not need a host for growth). The PCR-based method was deployed for identification of pathogen and later a phylogenetic analysis revealed the closest relationship to *Candidatus* phytoplasma (aster yellows subgroup 16SrI-B members). The application of these PCR-based approaches has improved our understanding of phytoplasma ecology and epidemiology (Martin et al. 2000). Real-time PCR, nested-PCR and microarrays were used to identify phytoplasma based on 16S rRNA gene sequences (Hogenhout et al. 2008). Further, whole genome sequencing of this deadly pathogen may provide insights to understand sandalwood-phytoplasm interaction leading to breeding for phytoplasm resistance.

10.2.3 Linkage Map Construction and Mapping

Estimation of recombination frequency between genes and assigning them on a chromosomes is a preliminary step of any gene/QTL mapping work. Such task is not yet explored in sandalwood. With the available genomic resources and genome-wide markers, linkage map construction is a accomplishable research task. Single tree genetic linkage mapping is one such option to explore in sandalwood which was done in other tree species (Tulsieram et al. 1992). SSR or SNP based high density linkage map construction are the other options to explore in sandalwood (Devey et al. 1996). GBS is emerging as most powerful tool for plant breeders in genome-wide association mapping and genetic linkage mapping (He et al. 2014). GWAS will unravel the complex genetic architecture and map the trait on the genome with high-throughput phenotyping data of the trait of interest (Hazzouri et al. 2019), whereas DArT, ddRAD and sequence capture are the other alternatives (Malebe et al. 2021; Chen et al. 2021; Zhang et al. 2020).

10.2.4 Marker Assisted Selection (MAS)

Identification and utilization of single nucleotide variation information from next generation technologies assist to overcome the drawbacks of conventional breeding in selection of saplings at early stage. Principal of MAS is exploring the linkage disequilibrium (LD) between markers and quantitative trait loci (QTLs) (Muranty et al. 2014). Implementation of MAS in crops to improve the traits are very much successful. However, MAS in tree species has shown limited success due to its complex inheritance pattern and out crossing nature, whereas markers having significant association from genome-wide association mapping can be utilized to overcome this drawback (Iwata et al. 2016; He et al. 2014).

10.2.5 Genomic Selection

Genomic Selection (GS) works on the principal of using genetic markers covering the whole genome and estimates the effects of all loci (Wang et al. 2018). GS in tree breeding is efficient in bringing the potential and radical improvement (Iwata et al. 2011). GS implementation on *Eucalyptus*, *Picea* and *Pinus* genera has shown significant results around the world. Establishment of breeding population holds primary importance for any plant breeding programmes, but breeding programmes of tree species are much more complicated and time consuming due to its perennial nature (Iwata et al. 2016). Whole genome re-sequencing (WGRS) can expedite with high marker density to cover the entire genome and make genome predictive models to be more accurate for the trait of interest, in addition to this WGRS can be utilized for capturing the allelic diversity and pan-genome analysis which will facilitate to gain the insight to evolutionary aspects and deep-level phylogenetic inference. Pan-genome analysis uses complete genome information and makes hidden genetic variation accessible to genetic studies of a given species (Jayakodi et al. 2020; Della Coletta et al. 2021). Conservation genomics is another booming platform used in identifying the genomic regions responding to selection and key genes of adaptive traits to environmental changes.

10.2.6 Metabolic Engineering

Sandalwood oil is highly priced for its unique fragrance. Terpene synthases (TPS) are responsible for converting the linear precursors like geranyl pyrophosphate (GPP), farnesyl pyrophosphate (FPP) and geranyl geranyl pyrophosphate (GGPP) into cyclic compounds. Cloning TPS and expressing in Bacteria or yeast system will help to produce santalol in a laboratory condition. Overexpression of TPS genes in sandalwood plant through a transgenic approach may cost the fitness of the plant. Excessive production of terpenoids in the plants will have considerable metabolic cost which tends to reduce the fitness of the plant (Gershenzon 1994). Understanding secondary metabolism pathways and regulation of genes in sandalwood should be a choice of research before venturing into transgenic production.

10.2.7 Genome Editing

Genome editing is recognized as one of the modern breeding tool. Sandalwood is also an important medicinal plant, rich source of antiseptic and astringent curing disease and ailments. Initial studies on alpha-santalol compound of sandalwood oil found to inhibit breast cancer cell growth in vitro by inducing apoptosis (Bommareddy et al. 2018), whereas intensive studies have to be conducted to understand the complete mechanism. Earlier reports of using CRISPR/Cas9 technology on mouse cancer model to identify novel tumour suppressor genes can be utilized for evaluating the alpha-santalol of sandalwood for its anticancer properties (Liu et al. 2017; Riordan et al. 2015). Attempts made towards understanding of biosynthesis of santalol yielded limited success (Niu et al. 2021; Zha et al. 2020), whereas genes involved in the biosynthesis of santalol and heartwood formation can also be a target study of CRISPR/Cas9. Knocking out undesirable genes in the sandalwood which contribute for contamination of essential oil fractions like farnesol is a need of the hour and currently farnesol being eliminated by fractional distillation. Farnesol is treated as potential allergen and editing farnesol synthase gene will be a boon to sandalwood oil industry. Usage of genome editing with CRISPR can also be a breeder choice in sandalwood breeding.

10.2.8 Supporting Farmers

Farmer's support system in terms of trainings and scientific inputs for sapling management are also lacking resulting in poor growth of sandalwood at commercial cultivation sites. Physical protection of mature trees is being the greatest problem, development of Remote surveillance systems are still under developmental stages. Interventions of Indian Government, Foreign Direct Investment (FDI) and NABARD are required in rendering the tax exemptions and financial assistance through Nationalized and commercial banks to support wide cultivation of sandalwood, whereas tree insurance kind of policy development must be taken care.

References

Bommareddy A, Knapp K, Nemeth A, Steigerwalt J, Landis T, Vanwert AL, Gorijavolu HP, Dwivedi C (2018) Alpha-Santalol, a component of sandalwood oil inhibits migration of breast cancer cells by targeting the β-catenin pathway. Anticancer Res 38 (8):4475–4480

Chen Y, Wu H, Yang W, Zhao W, Tong C (2021) Multivariate linear mixed model enhanced the power of identifying genome-wide association to poplar tree heights in a randomized complete block design. G3 11 (2):jkaa053

Della Coletta R, Qiu Y, Ou S, Hufford MB, Hirsch CN (2021) How the pan-genome is changing crop genomics and improvement. Genome Biol 22(1):1–19

Devey ME, Bell JC, Smith DN, Neale DB, Moran GF (1996) A genetic linkage map for Pinus radiata based on RFLP, RAPD, and microsatellite markers. Theor Appl Genet 92(6):673–679. https://doi.org/10.1007/BF00226088

Gershenzon J (1994) Metabolic costs of terpenoid accumulation in higher plants. J Chem Ecol 20(6):1281–1328

Hazzouri KM, Gros-Balthazard M, Flowers JM, Copetti D, Lemansour A, Lebrun M, Masmoudi K, Ferrand S, Dhar MI, Fresquez ZA (2019) Genome-wide association mapping of date palm fruit traits. Nature Commun 10(1):1–14

He J, Zhao X, Laroche A, Lu Z-X, Liu H, Li Z (2014) Genotyping-by-sequencing (GBS), an ultimate marker-assisted selection (MAS) tool to accelerate plant breeding. Front Plant Sci 5:484

Hogenhout SA, Oshima K, AMMAR ED, Kakizawa S, Kingdom HN, Namba S, (2008) Phytoplasmas: bacteria that manipulate plants and insects. Mol Plant Pathol 9(4):403–423

Iwata H, Hayashi T, Tsumura Y (2011) Prospects for genomic selection in conifer breeding: a simulation study of Cryptomeria japonica. Tree Genet Genom 7 (4):747–758

Iwata H, Minamikawa MF, Kajiya-Kanegae H, Ishimori M, Hayashi T (2016) Genomics-assisted breeding in fruit trees. Breed Sci 66(1):100–115

Jayakodi M, Padmarasu S, Haberer G, Bonthala VS, Gundlach H, Monat C, Lux T, Kamal N, Lang D, Himmelbach A (2020) The barley pan-genome reveals the hidden legacy of mutation breeding. Nature 588 (7837):284–289

Liu X, Wu S, Xu J, Sui C, Wei J (2017) Application of CRISPR/Cas9 in plant biology. Acta PharmaceutSin B 7(3):292–302

Mahesh HB, Subba P, Advani J, Shirke MD, Loganathan RM, Chandana SL, Shilpa S, Chatterjee O, Pinto SM, Prasad TSK (2018) Multi-omics driven assembly and annotation of the sandalwood (*Santalum album*) genome. Plant Physiol 176(4):2772–2788

Malebe M, Koech R, Mbanjo E, Kamunya S, Myburg A, Apostolides Z (2021) Construction of a DArT-seq marker–based genetic linkage map and identification of QTLs for yield in tea (*Camellia sinensis* (L.) O. Kuntze). Tree Genet Genom 17 (1):1–11

Martin RR, James D, Lévesque CA (2000) Impacts of molecular diagnostic technologies on plant disease management. Annu Rev Phytopathol 38(1):207–239

Muranty H, Jorge V, Bastien C, Lepoittevin C, Bouffier L, Sanchez L (2014) Potential for marker-assisted selection for forest tree breeding: lessons from 20 years of MAS in crops. Tree Genet Genom 10(6):1491–1510

Niu M, Yan H, Xiong Y, Zhang Y, Zhang X, Li Y, da Silva JAT, Ma G (2021) Cloning, characterization, and functional analysis of acetyl-CoA C-acetyltransferase and 3-hydroxy-3-methylglutaryl-CoA synthase genes in *Santalum album*. Sci Rep 11 (1):1–13

Riordan SM, Heruth DP, Zhang LQ, Ye SQ (2015) Application of CRISPR/Cas9 for biomedical discoveries. Cell Biosci 5(1):1–11

Tulsieram LK, Glaubitz JC, Kiss G, Carlson JE (1992) Single tree genetic linkage mapping in conifers using haploid DNA from megagametophytes. Bio/technology 10(6):686–690

Wang X, Xu Y, Hu Z, Xu C (2018) Genomic selection methods for crop improvement: current status and prospects. Crop J 6(4):330–340

Zha W, An T, Li T, Zhu J, Gao K, Sun Z, Xu W, Lin P, Zi J (2020) Reconstruction of the biosynthetic pathway of santalols under control of the GAL regulatory system in yeast. ACS Syn Biol 9(2):449–456

Zhang Q, Su Z, Guo Y, Zhang S, Jiang L, Wu R (2020) Genome-wide association studies of callus differentiation for the desert tree. Populus Euphratica. Tree Physiol 40(12):1762–1777

The manufacturer's authorised representative in the EU is Springer Nature Customer Service Centre GmbH, Europaplatz 3, 69115 Heidelberg, Germany. If you have any concerns regarding our products, please contact ProductSafety@springernature.com

Printed and bound by CPI Group (UK) Ltd, Croydon, CR0 4YY
30/09/2026
02225710-0011